T0297802

Assistive Technology
Service Delivery

Assistive Technology Service Delivery

A Practical Guide for Disability and Employment Professionals

Edited by

Anthony F. Shay, Ed.D., LPC, CRC

*Capacity Building Specialist, Assistive Technologist,
and Rehabilitation Specialist, University of Wisconsin-Stout Vocational
Rehabilitation Institute (SVRI), Menomonie, WI, United States*

Academic Press is an imprint of Elsevier
125 London Wall, London EC2Y 5AS, United Kingdom
525 B Street, Suite 1650, San Diego, CA 92101, United States
50 Hampshire Street, 5th Floor, Cambridge, MA 02139, United States
The Boulevard, Langford Lane, Kidlington, Oxford OX5 1GB, United Kingdom

Notices
Knowledge and best practice in this field are constantly changing. As new research and experience broaden
our understanding, changes in research methods, professional practices, or medical treatment may become
necessary.

Practitioners and researchers must always rely on their own experience and knowledge in evaluating and
using any information, methods, compounds, or experiments described herein. In using such information or
methods they should be mindful of their own safety and the safety of others, including parties for whom they
have a professional responsibility.

To the fullest extent of the law, neither the Publisher nor the authors, contributors, or editors, assume any
liability for any injury and/or damage to persons or property as a matter of products liability, negligence or
otherwise, or from any use or operation of any methods, products, instructions, or ideas contained in the
material herein.

British Library Cataloguing-in-Publication Data
A catalogue record for this book is available from the British Library

Library of Congress Cataloging-in-Publication Data
A catalog record for this book is available from the Library of Congress

ISBN: 978-0-12-812979-1

For Information on all Academic Press publications
visit our website at https://www.elsevier.com/books-and-journals

Working together
to grow libraries in
developing countries

www.elsevier.com • www.bookaid.org

Publisher: Mara Conner
Acquisition Editor: Fiona Geraghty
Editorial Project Manager: Lindsay Lawrence
Production Project Manager: Kamesh Ramajogi
Cover Designer: Matthew Limbert

Typeset by MPS Limited, Chennai, India

Contents

PART 1 THE ACCOMMODATIONS SYSTEM MODEL

PART 2 THE ASSISTIVE TECHNOLOGY SERVICE DELIVERY PROCESS

List of Contributors

Catherine A. Anderson
Rehabilitation Research and Training Center on Evidence-Based Practice, University of Wisconsin-Madison, Madison, WI, United States

Ray Grott
Rehabilitation Engineering and Assistive Technology (RET) Project, San Francisco State University (SFSU), San Francisco, CA, United States

Marcia Scherer
Institute for Matching Person and Technology, University of Rochester Medical Center; Physical Medicine and Rehabilitation and Senior Research Associate, International Center for Hearing and Speech Research, Webster, NY, United States

Anthony F. Shay
Capacity Building Specialist, Assistive Technologist, and Rehabilitation Specialist, University of Wisconsin-Stout Vocational Rehabilitation Institute (SVRI), Menomonie, WI, United States

Editor's Biography

Anthony F. Shay, Ed.D., LPC, CRC works in research and knowledge mobilization, capacity building, assistive technology, and rehabilitation with the University of Wisconsin, Stout Vocational Rehabilitation Institute. He holds a doctoral degree in education and a master's degree in counseling and psychological services. Dr. Shay is a licensed professional counselor, a certified rehabilitation counselor and holds an assistive technology and accessible design certification from the University of Wisconsin-Milwaukee. He has 30 years of experience working with individuals with disabilities in many contexts including extended employment, mental health counseling, state vocational rehabilitation, and in nonprofits. Dr. Shay currently works with Wisconsin's federally funded Assistive Technology Act Project, WisTech. He is chair of the Rehabilitation Engineering and Assistive Technology Society of North America's (RESNA) Vocational Rehabilitation Professional Specialty Group. Dr. Shay is a member of the WisLoan/Telework Boards (Wisconsin's assistive technology alternative loan programs), Wisconsin Rehabilitation Association, and chairs Everybody Works, a regional nonprofit in Southwest Wisconsin serving individuals with disabilities in employment.

Authors Biographies

Catherine A. Anderson, Ph.D., CRC serves as a researcher and director of the Rehabilitation Research and Training Center on Evidence-Based Practice at the University of Wisconsin-Madison. Educated as a rehabilitation counselor, Dr. Anderson worked in the field providing individualized job development and placement, evaluation and assessment, and work incentives benefits counseling services to consumers of both Vocational Rehabilitation and Medicaid-funded long-term support programs. More recently, her work has focused on disability and employment research with a specialized interest in studying and implementing knowledge transition strategies and processes for effectively moving research into practice. Dr. Anderson serves as principal investigator of the Program Evaluation and Quality Assurance, and Co-PI of the Targeted Communities Vocational Rehabilitation national technical assistance centers. She serves in a variety of state and national leadership capacities including the governor-appointed Wisconsin Rehabilitation Council and various national boards of directors. She actively collaborates with key stakeholders across systems including practitioners, administrators, researchers, service providers, individuals with disabilities and families in the areas of disability, employment, knowledge translation, vocational rehabilitation, poverty, and asset development.

Marcia Scherer, Ph.D. is a rehabilitation psychologist and founding President of the Institute for Matching Person & Technology. She is also professor of physical medicine and rehabilitation, University of Rochester Medical Center where she received both her Ph.D. and MPH degrees. She is the PM&R Department's research director. She is a past member of the National Advisory Board on Medical Rehabilitation Research, National Institutes of Health and is editor of the journal *Disability and Rehabilitation: Assistive Technology*. She is coeditor of the book series for CRC Press, Rehabilitation Science in Practice Series. Dr. Scherer is Fellow of the American Psychological Association, American Congress of Rehabilitation Medicine, and the RESNA. Dr. Scherer has authored, edited, or coedited 9 books and has published over 70 articles in peer-reviewed journals as well as 30 book chapters on disability and technology.

Ray Grott is director of the Rehabilitation Engineering and Assistive Technology (RET) Project at San Francisco State University (SFSU). He has a master's degree in rehabilitation technology from SFSU and is certified as an ATP (Assistive Technology Professional) and RET (Rehabilitation Engineering Technologist) from RESNA. Ray has worked directly with hundreds of people with a broad range of disabilities, providing assistive technology solutions. He is widely known for his creative and unique equipment modifications and custom design work for workers, students, and parents with functional limitations. He was coordinator of a federally funded graduate certificate training program in

RET at SFSU for 11 years and took classes in assistive technology for 20 years. Drawing on his extensive experience in assistive technology (AT) service delivery, he has presented numerous workshops and instructional courses at national conferences. Mr. Grott has been honored with both the Mentor and Fellow Awards from RESNA and his program earned the W.F. Faulkes Award from the National Rehabilitation Association for its "contribution of national importance to the increase of knowledge in the field of rehabilitation." He is currently serving as Chair of the Assistive Technology Advisory Committee to the California Department of Rehabilitation. He is also the immediate past president of RESNA.

Preface

In 2015, Anthony Shay became the first chair of the Rehabilitation Engineering and Assistive Technology Society of North America's Vocational Rehabilitation Professional Specialty Group (RESNA VR PSG). The first members of VR PSG cohort including John Bredehoft (Vice Chair), Ruth Carvalho, Ginny Gest, Shannon Aylesworth, and Bill Youngman with Robert Abegglen and Michael Lawler discussed the current issues in the field. A perennial concern regarding the interface between the vocational rehabilitation—assistive technology (AT) processes quickly became our focus giving rise to the VR PSG mission: "to advance knowledge transfer within the discipline of vocational rehabilitation toward increased VR professional assistive technology service delivery competence through technical assistance, communication, training and education" (Rehabilitation Engineering & Assistive Technology Society of North America, 2017). This book arose from an initial collaboration with Cayte Anderson at the University of Wisconsin—Stout Vocational Rehabilitation Institute regarding the challenges of knowledge translation between these two systems. The concept more fully developed through further collaboration with Ray Grott, and Marcia Scherer, all of whom generously offered their time and talents to writing this text.

For the professionals providing employment services to people with disabilities (disability-employment professionals), the AT world is a nebulous place and not well understood. Assistive technologies are integral in supporting competitive integrated employment outcomes, particularly for individuals with significant disabilities. The AT service delivery (ATSD) process fits well with the employment-related service delivery process. Yet, the ATSD process remains mysterious for many of the professionals practicing in the field. In fact, disability and employment professionals and AT professionals tend to be unsure of where these two systems interface and how to effectively bridge the gap between them. Translating this knowledge into practical information to enhance understanding of these systems by practitioners and stakeholders, and encourage application in practice, serves to improve the lives of people with disabilities through meaningful and satisfactory employment outcomes.

It is our sincere hope that through the delineation of the accommodation system and the ATSD process, practitioners across varied fields carry this knowledge into the contexts in which they serve and are able to more effectively address the needs of the people who rely on them.

REFERENCE

Rehabilitation Engineering & Assistive Technology Society of North America. (2017). *Vocational Rehabilitation Professional Specialty Group*. Available at: ⟨http://www.resna.org/professional-development/volunteer-and-leadership-opportunities/special-interest-groups/professional⟩.

Acknowledgments

We very much appreciated the passion and enthusiasm received from Priscilla Matthews and Melissa Lemke as we began this project. We also want to thank Johan Borg for his contributions to this effort. We recognize the patience of our family and friends (especially Jacqueline Shay). Dr. Richard E. Morehouse, you are a true friend and colleague. As you indicated many years ago — hope springs.

This book is dedicated to all the idea technologists transforming inert pools of information into wellsprings of application.

The accommodations system model

WHAT IS THE ACCOMMODATION SYSTEM MODEL?

Talking about effective task engagement at home, work, in school, for sports, or to do any kind of activity can be difficult. When we need accommodations to do this, it can be even more difficult. The accommodation system (AS) model helps us organize our thoughts around basic need areas. We consider all the areas which give rise to the need. Determining need goes much further than simply deciding what assistive technologies to use. There are many things to consider. Even when we find what seems like a very good match to an individual's needs, the technology may still go unused. We must consider the person, the environment, the task engagement activity, social integration, effective accommodations and how all these elements come together. We call this our situated experience.

WHY IS THE ACCOMMODATION SYSTEM MODEL IMPORTANT?

Being employed is one of the primary ways in which we find meaning in life (Haworth, 1997). As a result, we need a better understanding of how to obtain, maintain, or improve our work situation when we are or become disabled. We can

- use the AS model as a guide;
- work toward our goals without giving up;
- anticipate and plan for as many needs as possible;
- avoid spending money on things we do not need or will not use;
- consider how other people will see us and how we will feel when we use accommodations; and
- overall, we can consider whether we will be satisfied with how the accommodations should help us on the job.

Because we are all different, the AS model may also offer us insight into areas we may not have otherwise considered.

WHAT IS THE STRUCTURE OF THE ACCOMMODATION SYSTEM MODEL?

The AS comprises five core domains: Self, Others, Environment, Accommodations, and Situated Experience. The domains are placed in a large inner circle within a circle (an inner circle within an outer ring). The self is in the inner circle at the center of the model. The outer ring is divided into four equal parts each of which represents Others, Environment, Accommodations, and Situated Experience. Each of these domains is separated by a broken line between the elements. This illustrates how closely each domain interacts with each of the others. The Self is in a large circle at the center of the model to illustrate that people are the central focus in the AS. The domains in the outer ring (i.e., Others, Accommodations, Environment, and Situated Experience) interact with each other but all find meaning in the Self. In other words, each of these domains is important, but we as individuals make sense of them. We are working to achieve our goals when these domains are aligned. We find satisfaction and meaning in our work lives. See Fig. 1A for an illustration of the accommodation system model.

ACCOMMODATIONS AS ACTIVITY ENABLERS

The enablement model is another way to view the AS model. By using accommodations, we are able to work to the greatest extent possible (Stephens, 2009). As a result, we can more fully participate in community life. This gives us a sense of hope, self-reliance, and true enablement (Elliott, Kurylo, & Rivera, 2005). With true enablement, or true rehabilitation, we are able to

- be as independent as possible in all aspects of our lives throughout our lifespan;

FIGURE 1 (A)

The accommodation system model.

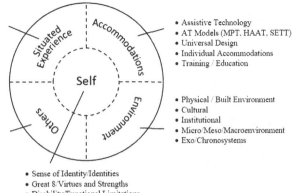

- Self-Efficacy/Vocational Competence
- Level of Stimulation/Arousal
- Affective State
- Motivational Quality of the Experience
- Satisfaction and Meaning

- Interpersonal Connectedness
- Social Competence
- Interpersonal Orientation
- Self/Other Awareness
- Identity/Social Validation

- Assistive Technology
- AT Models (MPT, HAAT, SETT)
- Universal Design
- Individual Accommodations
- Training / Education

- Physical / Built Environment
- Cultural
- Institutional
- Micro/Meso/Macroenvironment
- Exo/Chronosystems

- Sense of Identity/Identities
- Great 8/Virtues and Strengths
- Disability/Functional Limitations
- Vocational Competence
- Cultural Competence

FIGURE 1 (B)

Elements of the accommodation system model.

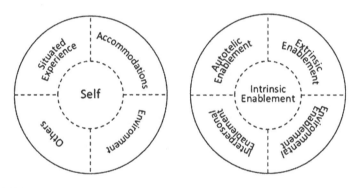

FIGURE 1 (C)

The accommodation system and enablement domains.

- improve the quality of our lives;
- be connected with others in a meaningful way;
- better understand our accommodation needs, how they may change, and how well they have met our needs.

We can expect accommodations to facilitate practical outcomes allowing us to better navigate the environments and engage in the tasks that are expected of us in employment (Stephens, 2009).

We present the Accommodation and Enablement models side-by-side for an easy comparison (see Fig. 1B): each domain of the AS (on the left) reflects the corresponding enablement domain (on the right): Self (intrinsic enablement),

Others (interpersonal enablement), Environment (environmental enablement), Accommodations (extrinsic enablement), and Situated Experience (autotelic enablement). We discuss each of these domains in the following chapters (Fig. 1C).

REFERENCES

Elliott, T. R., Kurylo, M., & Rivera, P. (2005). Positive growth following acquired physical disability. In C. R. Snyder, & S. J. Lopez (Eds.), *Handbook of positive psychology* (pp. 687−699). New York, NY: Oxford University Press.

Haworth, J. T. (1997). *Work, leisure, and well-being*. London: Routledge.

Stephens, D. (2009). *Living with hearing difficulties: the process of enablement*. West Sussex, England: John Wiley & Sons.

Accommodation System: Self

1

Anthony Shay

Capacity Building Specialist, Assistive Technologist, and Rehabilitation Specialist, University of Wisconsin-Stout Vocational Rehabilitation Institute (SVRI), Menomonie, WI, United States

WHAT IS THE SELF?

The *self* is who we are. It is all the individual pieces by which we recognize ourselves as being different and yet the same as everyone else. Our sense of self or "I" is dependent upon an "elemental consciousness" we derive from an "*immediacy*" and "'*certainty*' of experience" (emphasis in the original) (Erikson, 1997, p. 86). The Self is central to the accommodation system (AS) model primarily due to the central nature of the "self-observing I" and its fundamental importance to the study and understanding of the interplay and ramifications of task engagement on the other AS domains (Erikson, 1997, p. 87). It is these aspects of self which must be accounted for when we decide to find or change employment. The *self* is the first of the five domains of the AS we will be discussing in Part 1 of this text. Each of these domains (i.e., Self, Others, Environment, Accommodations, and Situated Experience) interacts with and influences the others. The AS model offers a framework for understanding the role of the self in relation to other core domains of Other, Environment, Accommodations, and Situated Experience in the provision of accommodations in employment.

THE SELF IN THE CONTEXT OF EMPLOYMENT

The more we invest of ourselves in the job development process the more meaning and satisfaction we derive from the subsequent job we find. We can define satisfying and meaningful work as such when it reflects our primary employment factors: strengths, resources, priorities, concerns, abilities, capabilities, interests, and informed choice (Giesen & Hierholzer, 2016, p. 175; Vocational Rehabilitation, 2016) or, in other words, our Great 8. Using the Great 8 as a guide we ensure we take an individualized, person-centered approach to making employment-related decisions which leads to job satisfaction (Cobigo, Lachapelle, & Morin, 2010). These decisions are also

Assistive Technology Service Delivery. DOI: https://doi.org/10.1016/B978-0-12-812979-1.00001-1

influenced by mental processes both of which we are aware and of those of which we are not fully aware (Feuerstein, Feuerstein, & Falik, 2010, p. xvi).

THE GREAT 8 IN MORE DETAIL

The Great 8 represents aspects of the self which are helpful considerations in making work-related decisions which allows us to effectively compete in the job market leading to gainful employment. Factors such as these are especially useful in developing employment goals and defining a career or job orientation. Although each of us are different—we each bring a unique mix of these qualities to employment—it is the quality of our situated experiences that matter most (Wolfe, 2001). We are complex, dynamic, and unique. As a result, we not only find greater meaning in jobs that are closely matched to the attributes of the self, the jobs we obtain last longer (Bond, 2004). We define the Great 8 is as follows:

1. **Strengths:** Positive qualities or innate proficiencies. Strengths may include attributes such as intellect, physical ability (e.g., motor coordination, task speed, and accuracy), language mastery, self-regulation, intrinsic motivation, and talents (e.g., being mechanically inclined). Strengths may be developed through paid work experience, volunteer experience, hobbies, and other life experiences. Strengths are virtues which are universal, essential, and an irreducible part of our character (Peterson, 2006). Virtues and strengths are discussed in more detail later in the chapter.

2. **Resources:** Resources refer to the supports we have available to us (e.g., job coaching, food pantry, support groups) toward achievement of an employment goal. They may be personal, community, or other more peripheral supports. Individual resource mapping allows us to visualize and plan for resource acquisition and utilization as we prepare for employment. Maps illustrate the range of supports to which we may have access. Resources, such as transportation, may be provided by more than one source. An example might be when we borrow a friend's vehicle to go to a volunteer work experience and get money from our parents for gasoline, stop by the service station to put air in the tires, and hang a disability placard in the window so we can park nearer to the building on the work site. We typically seek resources beginning with those we personally possess or to which we have direct access and then move out toward those that might take some time and effort to obtain. Resources can also be developed specifically to address disability and employment needs. When we receive resources from an employer as a component of a job, they are considered natural supports (Minor & Bates, 1997; Peterson, 1995).

   When we map resources, working outward from the center of the individual resource map, (see Fig. 1.1), we may consider

   a. Resources to which we have direct access and personal control of including personal finances, a personal vehicle, and bank accounts;

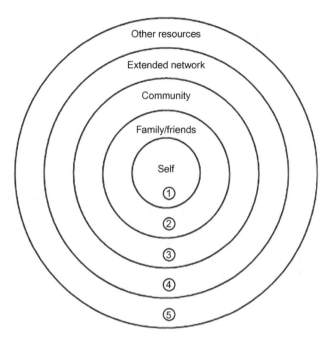

FIGURE 1.1

Resource mapping.

b. Family and friends who may provide emotional and financial support, personal care services, or companionship;

c. Community services such as parenting, disability, or alcohol support groups or acquaintances;

d. Governmental agencies or nonprofit organizations may offer rehabilitation or financial assistance programs;

e. We may need to network and seek out resources which may be unknown to us, may require a referral (or an invitation from a member), or those that may be difficult to access (i.e., eligibility-based or programs/services which are membership restricted based on established criteria and standard operating procedures).

As we move further away from the center, we have less control and influence over the resources we are seeking. Our situated experience can impact access to resources. Chapter 5, Accommodation System: Situated Experience goes into more detail regarding situated experiences.

As a job development tool, resource mapping provides a means by which we can consider interest area(s) for employment and the associated geographic locations mapping our resource needs and their availability in those locations. We must consider our home (or prospective home) and all employment contexts as centers of a job search target area keeping in mind access to the resources we need. This can facilitate a more proactive and practical job search. Identifying and planning access to necessary resources can facilitate job maintenance by reducing the number of barriers to keeping a job or minimizing their impact on job maintenance.

3. **Priorities:** Priorities are concerned with the recognition of a rank, order, or of items given special status over others. We assign value to an item based on this recognition which affords it special status. Priorities are a reflection of the grounding of competing values as they relate to our personal well-being and the availability of alternative options. Positive personal experiences are further reinforced when our priorities, goals, and the context of a goal attainment activity are aligned (Sagiv & Schwartz, 2000).

4. **Concerns:** Concerns, similar to interests and priorities, are a recognition of the value we place on items relevant to goal attainment activity, context, perception, and changes that happen over time. Concerns imply that we are motivated to act in certain ways. We focus our attention in a deliberate way when we engage in work activity. We prioritize these actions based on the values we hold as they relate to our personal well-being (Kingsbury & Carr, 1939).

5. **Abilities:** Like strengths, abilities may involve the use of attributes such as intellect, physical ability, language mastery, self-regulation, intrinsic motivation, and talents. They may also be developed through work experiences, volunteer experience, hobbies, and other life experiences. Our abilities inhere in natural aptitude or develop through acquired proficiency (National Consortium of Interpreter Education Centers, 2017). Abilities tend to be enduring and context-dependent traits we are born with and develop over time. Abilities are predictive in the sense that we can expect to develop and repeat them over time (Dawis & Lofquist, 1984). We also tend to label abilities by observation. In other words, we attribute our abilities to the context or actual task or behavior from which the label arises (i.e., being a good ball player or having mechanical aptitude) (Josselson, 1992). The term aptitude tends to be used synonymously with ability. Aptitude refers to the nature of possessing a particular ability. It most typically is used as a measure for potential self-efficacy in task engagement and in employment capabilities (Carr & Kingsbury, 1938; Dawis, 1996).

6. **Capabilities:** Capabilities are essentially aspects of a belief system we are constantly redefining based on how well we think we will do while engaged in a given task or activity. The term self-efficacy tends to be used synonymously with capability. Four types of experiences relate to how self-efficacy impacts task-oriented belief systems which represent keys to successfully building capabilities:

   a. The first experience type relates to the tried and true maxim "if at first you don't succeed, try, try again." Failure and learning to overcome challenges are keys to building capabilities.

   b. The second experience type relates to vicarious experiences which build capabilities and vocational competence. This is a reflection of how closely we identify with the models of the behavior we are observing based on their relative success and failure in the tasks observed. This builds belief in our own capacity to perform the same or similar tasks (Bandura, 1997; Erikson, 1980).

c. The third experience type involves social influence. When another individual, such as a job coach or manager, provides verbal encouragement toward on-task behavior, it tends to reinforce continued task-related activities through increased engagement and a more sustained effort (Bandura, 1997; Erikson, 1980).

d. The fourth experience type is affective valence (i.e., how we feel we are doing): When we have positive experiences, we want to continue to work, and when we have negative experiences, we do not want to work (Bandura, 1986, 1991, 1997). We also use the term propensity to describe experience-based capacity to learn. Propensity refers to how ready, willing, and able we are to engage in and take advantage of learning opportunities (Feuerstein, Falik, & Feuerstein, 2015, p. 30).

7. **Vocational Interests:** Interest are indicators of how invested we are in a particular goal (Kanfer & Ackerman, 2005). They are specific, dynamic, and a natural part of the thought process (Dewey, 1963, p. 195). Interests *are* emotions. We give expression to them through verbal and nonverbal behavior (e.g., vocal inflection or raised eyebrows). Interests form the foundation for how we think, feel, and act about things (Sylvia, 2008, p. 57). This is a fundamental part of how motivated we are to work. When we have an interest in activities, we choose them for the sake of doing them more so than for extrinsic rewards. Kingsbury and Carr find interests as coming from personal experience imbued by variables within and outside of us (e.g., the other elements of the Great 8). Our vocational interests come from these variables (Kingsbury and Carr, 1939, p. 203−204).

   Abilities and interests are closely related. In fact, our interests take form early in life and become associated with skills and abilities over time. This may be why interest tends to endure for long periods of time (Kanfer & Ackerman, 2005). Like abilities, interests tend to be labeled in terms of the contexts within which they arise (Kingsbury & Carr 1939, p. 203−204; Sylvia, 2008). Additionally, when we refer to an interest, we also refer to the intensity of the expected goal striving behavior. Intensity is based on our perceived need for achievement (e.g., competitiveness, peer pressure, we want to change how we feel). The value we place on task engagement drives the intensity of our interest in the activity (Kanfer & Ackerman, 2005; Lewin, 1999).

8. **Informed choice:** Informed choice is a decision-making process whereby a person with a disability can fully participate in the employment process and make meaningful decisions. Full participation means we are given all of the relevant information necessary to make employment-related decisions. The maxim "nothing about me without me" reflects the notion of informed choice. People with disabilities and/or their legal guardians make these decisions

together with the professionals providing them disability and employment-related services.

Choices are based on what is reasonable, necessary, and appropriate in service of disability and vocational goals rather than taking a random approach or basing decisions strictly on wants and "nice-to-haves." We may have disability and employment focused teams comprised of professionals from a variety of agencies or systems with whom we are working. We encourage and consider their input, but in the end, we make the final decision. Informed choice also involves communicating with others using the most appropriate mode of communication and providing access to necessary accommodations; using the preferred mode of communication of the individual with the disability. This helps facilitate full participation in the process. Exercising informed choice is a microcosm of the overarching goal of employment and disability services seeking to empower and build self-reliance in the people they serve (Kosciulek, 2007; National Consortium of Interpreter Education Centers, 2017; Wolff-Branigin, Daeschlein, Cardinal, & Twiss, 2000).

CONSIDERING VIRTUES AND STRENGTHS IN THE EMPLOYMENT CONTEXT

Professionals in the field of disability—employment consider, among other things, the strengths an individual brings to employment within the context of service delivery. However, few of them can provide an effective description of a strength. The following discussion addresses consumer strengths in the context of disability-employment services toward greater clarity. Peterson (2006) offers a model which provides some clarity.

Peterson's model, the *Virtues in Action-Inventory of Strengths* (VIA-IS), was developed as a companion to the American Psychiatric Association's *Diagnostic and Statistical Manual of Mental Disorders* (DSM) which he considers the epitome of the medical model applied to psychological disorders. As a companion, the VIA-IS provides an alternate image of disorders which focuses on issues which could be defined as functional limitations rather than disorders making it readily adaptable to the disability—employment field.

The DSM relies heavily on individual self-report to diagnose disorders. Peterson argues that self-report is also a viable way to diagnose strengths. His model does not take the traditional what-is-wrong-that-needs-to-be-fixed approach to human well-being. Rather, he takes a what-is-right-that-can-be-improved-upon approach. Virtue deficiencies or absences lead to distress which may be misdiagnosed as psychological disorders. To address these issues, Peterson devised a taxonomy of human strengths which boasts substantial support across research literature from philosophy to evidence-based practice and across time from Aristotelian works to cognitive therapy tenets.

According to Peterson, strengths are comprised of six virtues which, in turn, are comprised of several character strengths. The virtues are the same as those recognized in antiquity through modern times. Virtues have a lasting appeal. Observing a virtuous life can serve as an indicator exemplifying a sought-after identity (Csikszentmihalyi & Csikszentmihalyi, 1988). To the extent that they reflect identity, we assign value to them. These virtues represent the universal, essential, and irreducible values that have ensured the survival of the human race: wisdom, courage, love, justice, temperance, and transcendence. As disability—employment professionals who help others find, maintain, and improve jobs can attest, it can be exceedingly difficult to differentiate between abilities or talents, and strengths or virtues. Can we answer the questions "what are you *able* to do"; "what are you *capable* of doing"; and "what are your strengths" without stumbling over our answer? Peterson offers this insight: "talents and abilities can be squandered, but strengths and virtues cannot"—virtues and character strengths are highly valued in their own right and not sought after for extrinsic reasons. Peterson and Seligman further qualify this stating, "strengths have moral value and are acquired and developed dynamically" (Littman-Ovadia, Lazar-Butbul, & Benjamin, 2014).

Each of the six virtues is comprised of several character strengths. Character strengths are the mechanisms by which the virtues manifest themselves and are most often those qualities referred to as personal strengths. Table 1.1 outlines the relationship between virtues and character strengths.

Peterson (2006) argues that the VIA-IS may be an effective means to address issues through better differential diagnosis. Diagnoses may be enhanced when considering consumer character strengths, quite possibly through the lens of medical issues and psychosocial stressors, which may manifest as an absence or exaggeration of character strengths. A better understanding of our strengths (and consequently exaggerations or absences) provides another perspective by which assistive technology and disability—employment professionals may identify more effectively the functional limitations that negatively impact our vocational success. Additionally, Peterson claims that the exaggeration of character strengths is less often a problem than is its absence. Addressing issues related to shoring up character strengths of an individual (e.g., counseling and accommodations) may serve to buttress other efforts toward meeting their needs and achieving programmatic goals—increasing vocational competence, rates of employment, and job development resiliency (Littman-Ovadia et al., 2014).

FORMING A VOCATIONAL IDENTITY

The process of striving for a vocational identity (e.g., registered nurse, brick layer, customer service representative) is dynamic and complex. We begin to develop a sense of identity very early in childhood with development continuing throughout our lives. As we learn more about ourselves and the world around us,

Table 1.1 The Vocational Application of Virtues and Strengths

Functional Area	Character Strength	Virtue	Vocational Application
		Wisdom & Knowledge	
cognitive strengths: acquisition of knowledge	1. creativity 2. curiosity & interest in the world 3. judgment & critical thinking 4. love of learning 5. perspective		1. novelty and generativity 2. knowledge seeking, exploration, and discovery 3. objectivity, reflection 4. personal development and skill mastery 5. rationality and providing counsel
		Courage	
emotional strengths: exercise of will / overcoming adversity / goal focused	1. bravery 2. persistence 3. authenticity / honesty 4. vitality		1. facing challenges & opposition, acting on principal 2. performing functional task elements 3. responsibility & accountability, without pretense 4. doing things wholeheartedly, with excitement & energy
		Love	
interpersonal strengths: being of service / a friend to others	1. intimacy 2. kindness 3. social intelligence		1. closeness to others, reciprocating caring, and sharing 2. being of service to others 3. emotional resonance with others and being social adept
		Justice	
civic strengths: healthy community life	1. citizenship 2. fairness 3. leadership		1. teamwork, loyalty, doing your share of the work 2. unbiased and fair decision-making 3. facilitating/organizing group process and outcomes
		Temperance	
intrapersonal strengths: exercising self-control	1. forgiveness/mercy 2. humility/modesty 3. prudence 4. self-regulation		1. avoiding vengefulness, giving second-chances 2. avoiding self-aggrandizement or promotion 3. observing discretion and avoiding risks 4. intrinsic control over thoughts, feelings, and behavior.
		Transcendence	
existential strengths: making connections to the larger universe & providing meaning	1. appreciation of beauty and excellence 2. gratitude 3. hope 4. humor 5. spirituality		1. recognition of beauty and excellence in all areas of life 2. awareness and ability to express gratitude 3. expectation of and work toward a good future 4. seeing the lighter side of things and expressions related to this 5. having a coherent belief in a higher purpose or meaning and knowing where you fit in this scheme—shapes behavior and provides comfort

we begin defining vocational aspirations. Disability and functional limitations can have a serious impact on vocational identity—whether we are just beginning a foray into the world of work, trying to maintain or looking for new job opportunities. Identity development tends to be an overlooked aspect of vocational goal striving activity. Accommodations such as assistive technology can enable identity striving behavior and even reinvigorate the employment process (Shay, Anderson, & Matthews, 2017).

Motivation for the development of vocational identity may be intrinsic, extrinsic, or a combination of both. It may follow formal or informal means (e.g., highly structured like a university degree or more passive such as through vicariously observing someone performing a job from afar). Context is essential to identity development (Kasl & Elias, 2000). We can learn much through immersing ourselves into the work environments in which our prospective occupational interests are acted out. We learn from others who do the work in which we have an interest. In fact, social interaction validates and reinforces vocational goal-striving behavior. We look to align ourselves with and learn from others who already hold the identity toward which we strive. We also seek recognition and appreciation from them and others who populate our social networks (Gollwitzer & Kirchhof, 1998).

Building a vocational identity involves gathering the identity indicators (i.e., symbols) which reflect the vocational identity we seek. These may be highly challenging to attain such as a graduate degree or a state licensure. More easily attained indicators may include work-related books, tools, or clothing (Wicklund & Gollwitzer, 1982). Obtaining employment symbols is central to the process of gradually moving along our vocational learning trajectory. When we frame employment goal striving as vocational identity development and provide effective accommodations that allow for access to the indicators and symbols which define identity, we provide people with disabilities seeking employment an opportunity to develop "a strong identity, engage in more effective goal directed behavior, and find greater vocational satisfaction" (Shay et al., 2017).

Lave and Wenger described the process of building an identity, expertise, and a sense of belonging in a community of practice as a product of job training. This training reflects "engagement in social practice that entails learning as an integral constituent" (Lave and Wenger, 1991, p. 35). When our learning trajectory is borne of genuine access to a vocational community, we build a vocational identity (Fenton-O'Creevy, Dimitriadis, & Scobie, 2015; Kasl & Elias, 2000) characterized by greater task mastery (i.e., value-in-practice) across the communities and landscapes in which we practice—within a particular job and those associated with it (Fenton-O'Creevy et al., 2015).

As we get better at the work we do, we are more likely to want to continue doing it (Bandura, 1971). Receiving immediate feedback on our performance allows us to more effectively work through the challenges that we face on the job. As we get better and expect to do well on our jobs, we are empowered to make the changes we see as necessary to achieve our goals. People who seek to optimize on-task behavior like this have self-motivating personalities. They identify with the work and tend to engage in activities for the sake of doing them because they find them highly satisfying. Attitudes like this drive vocational competence-building. Virtually any activity can be self-motivating and lead to increased competence (Csikszentmihalyi, 1975).

DEFINING VOCATIONAL COMPETENCE

The core features of competence have been described in many ways. As facets of the self, these are innate qualities which may be stable (e.g., self-concept) or fluctuate (e.g., self-efficacy). They may reflect wide-ranging levels: low, medium, high (e.g., self-confidence) or qualities such as tasks, time, and context-dependent experiences (e.g., self-efficacy) (Csikszentmihalyi, Abuhamdeh, & Nakamura, 2005; Day & Jutai, 1996; Deci & Moller, 2005; Kanfer & Ackerman, 2005). When we build competence, we are moving through a process whereby we set goals for ourselves, strive to achieve these goals, and evaluate progress toward the goals. Based on self-appraisal, we establish meaning and satisfaction derived from task engagement activity (Langer & Dweck, 1973). Cook and Polgar (2012) and Bridges (1993) offer perspectives on the concept of vocational competence.

The Cook and Polgar (2012, p. 24) model of occupational competence is comprised of five elements including capacity—potential application of knowledge, skills, and abilities; effectance—the actual application of knowledge, skills, and abilities within a task; affordances—perceived extrinsic enablers of a task; self-efficacy—belief in one's ability to realize success in a task; and competence—realization of successfully endeavoring toward task completion according to a set norm.

Bridges (1993) conceptualizes metacompetence through a transferable skills analysis framework. He proposes a first and second order skill set and a first and second order competence set. The first-order skills are the situational, transferable skills. The second order skills or the metaskills are those skills which allow for the effective development of the first-order, transferable skills. Similarly, first-order competence (or the situational competence) allows for efficacy in use of transferable skills. Second order competence (or meta-competence) allows for the development and use of first-order competencies: "The person with metacompetence has, as it were, a bird's eye view of the particular competence (even while doing it) which allows them to recognize that it depends on a conjuncture of circumstances that can and probably will change." Bridges (1993, p. 9) goes on to say that cognitive skills, social skills, and goal orientation all serve this metafunction, particularly given its application to novel tasks. Fig. 1.2 illustrates the relationship between first and second-order competence and first and second-order skills.

Our perception of vocational competence hinges not on the knowledge, skills, and abilities we develop for use in work activity. Rather, task efficacy lays in the belief that we are able to perform expected work tasks and do so consistently across multiple contexts (Bandura, 1997).

Our vocational interests lay the foundation for competence. We begin to build competence when we identify with and begin striving toward an employment goal. Having options drives learning and goal striving and builds competence (Meyer, Rose, & Gordon, 2014). The degree to which we succeed at this, our level of goal striving success, tends to drive continued interest (Bruner, 1966,

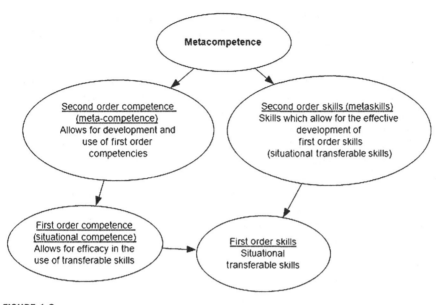

FIGURE 1.2

Metaskills and metacompetence.

p. 118; Meyer et al., 2014). It is "directive, selective, and persistent." We choose to continue to engage in the work activity because it provides us a sense of satisfaction (R. W. White in Bruner, 1966, p. 118). We build value-in-practice and a vocational identity with time spent on-task. We can characterize this as movement through a stepwise learning process: beginner, intermediate, and expert practitioner (Lave & Wenger, 1991). Each step along this continuum reflects an increase in our base of vocational knowledge and the skill with which we select relevant task knowledge and apply it within specific work contexts (National Research Council, 2000). Each step in this process reflects greater self-confidence and self-efficacy (Ajzen, 1991). However, disability can create challenges with effective task engagement in the workplace jeopardizing vocational competence.

A PRIMER ON DISABILITY

Human functioning is comprised of three primary components: (1) body/body part, (2) the whole person, and (3) the entire person within a context (World Health Organization, 2001). How the self functions speaks to our mechanical, actional, and cognitive properties. We function in time/space, are goal oriented, and are cocreators of meaning. We call this *agency* (Leslie, 1995). Our sense of agency is drawn from intention, reinforced by behavior, and may require an interdependent effort of the self with others (Bandura, 2001). Disability limits human functioning as the "result of a complex relationship between an individual's health condition and personal factors, and of the external factors that represent the circumstances in which the individual lives" (World Health Organization, 2001).

According to Prince, an integrative international perspective of disability includes "body structures and functions, daily activities, and social participation ... and the role of environmental factors" resulting from a "physical or mental condition or health problem" that may or may not be visible to others and has a limiting impact on their activities (Prince, 2017, pp. 76–77). Disability creates "a gap between a person and his or her environment" (Adya, Samant, Scherer, Killeen, & Morris, 2012, p. 77; Jette, 2006, p. 728). A disabling condition is anything that gets in the way of the self and the performance of activities in the environments in which they occur. Lange describes disabling conditions in terms of disability, impairment, activity limitation, and participation limitation and offers an excellent example to illustrate the distinctions between them:

- Disability: "a client has a spinal cord injury";
- Impairment: paralysis;
- Activity limitation: "may be independent use of public transportation";
- Participation restriction: "may be a [subsequent] lack of ... ability to participate in church" (Lange, 2008, p. 1).

Possessing a disabling condition is as much a product of others' perceptions as it is actually having the condition.

Kasnitz, Switzer Fellow, and Shuttleworth (2001, paragraph 2) find that people with disabilities have an impairment when they possess actual or perceived "problems, illnesses, conditions, disorders, syndromes, or other similarly negatively valued differences, distinctions, or characteristics [resulting in] ... situational views of cause and cure and of fate and fault." In other words, we have a disability when we have or are treated as though we have a disabling condition. The concept of disability is complex and covers a lot of semantic ground including (but not limited to) the following variables:

- Having a disabling condition;
- Etiology, prognosis, chronicity, and severity of a disabling condition;
- Physical/psychological adjustment to a disabling condition;
- The perception of having a disabling condition;
- Being associated with someone who has a disability;
- Hidden versus obvious disabilities;
- Physical versus psychological disabilities;
- Societal implications;
- Requiring accommodations;
- Whether a functional limitation or impairment is present and the impact it has on specific activities within specific contexts;
- Access to effective accommodations
- Social acceptance/stigma;

DISABILITY MODELS

Disability models are designed to provide a means to better understand what is meant when the term disability is used. Models can address what needs to be "fixed" or they can focus on and reinforce individual empowerment and positivity, a concept also known as "principled optimism" (Tilson & Simonsen, 2013) and well known to the fields of education and positive psychology. We provide a brief description of several types of models below (Chan, Sasson Gelman, Ditchman, Kim, & Chiu, 2009). Our objective is not to offer a comprehensive or mutually exclusive list, but rather, an introduction to a variety of perspectives. Of note is the growing support for a universally recognized model of disability and functional limitations. The World Health Organization, International Classification of Functioning, Disability, & Health (ICF) appears to be the front-runner as the most applicable model thus far (Federici, Meloni, & Corradi, 2012).

Medical/biomedical model: Focuses on the biological causes of illness and disability. This is a curative model; something causes the illness or disability and it must be resolved, accommodated, or functioning restored.

Functional Model: Delineates disabling conditions in terms of "pathology, impairment, functional limitation, and disability" (Chan et al., 2009). This model is concerned with the self in context and how the disabling condition impacts functional abilities.

Universal/biopsychosocial model: This model adds social, psychological, and behavioral elements to the biomedical model. It looks at the interrelationship of disabilities inherent in an individual and expressed across the population as a whole and whose meaning is negotiated as a universal experience.

Biomechanical model: Focuses on modeling and testing prior to development and implementation of interventions. The practice of using representations for actual systems (physical or virtual) are used to measure and test the mechanical properties of biological systems under simulated conditions.

Social Model: Social models focus on environmental, interpersonal, and functional limitations. Enabling function, overcoming functional deficits, and full community integration are also emphases of this model (Chan et al., 2009).

DSM-5: Employs a nonaxial system for diagnosis, treatment, and prognosis of psychological functioning. DSM-5 does adopt some features from ICF (psychosocial and contextual problems as they relate to mental functioning). Previous versions of the DSM used an axial diagnostic scale including a Global Assessment of Functioning—a subjective measure of overall functioning (American Psychiatric Association, 2013).

World Health Organization, International Classification of Functioning, Disability, and Health (ICF): Focuses on health and disability problems at the micro (individual) and macro (population) levels including contextual considerations. "The ICF identifies 3 levels of human function: functioning at the level of body parts, the whole person, and the whole person in their complete environment. These levels in turn, contain three domains of human function: body functions and structures, activities, and participation" (Jette, 2006, p. 733; World Health Organization, 2001).

Positive psychology: Focuses on human virtue and strengths, looking beyond disability to ability. This approach defines satisfaction and meaning in life as being derived from building on individual strengths and how this can be improved-upon and reinforced in the individual.

NOTE: Frailty, typically discussed in reference to the aging population, is correlated with disability and the simultaneous occurrence of more than one disabling condition. However, frailty tends to refer to "progressive weakness and vulnerability" which may cooccur and influence one or more disabling conditions. "The causal interconnectedness" of disability, frailty, and cooccurring disabling conditions can make identifying the occurrences (or cooccurrences) of disability difficult to diagnose (Fried, Ferrucci, Darer, Williamson, & Anderson, 2004, p. 261).

FUNCTIONAL LIMITATIONS

Understanding functional limitations and the challenges associated with a return to work can help facilitate task analysis, determining accommodation strategies, and in facilitating job maintenance. Three tests can determine if a work-related functional limitation exists: First, is the limitation a direct result of a disability? Second, does the disability have an impact on the individual's ability to engage in work-related activity? Third, when compared to the workforce in general, can an average member of the workforce do (or not do) the work activity as compared to the individual with the functional limitation? If the answer to these questions is yes, a functional limitation likely exists. Other factors may influence whether a disabling condition results in a work-related functional limitation. Not being able to drive due to a medical restriction due to a seizure disorder would be a functional limitation in mobility due to disability. Not driving because you do not have a driver's license because you have not yet taken the driver licensing exam would not. An assessment of functional limitations also takes into consideration whether there are accommodations in place. With effective accommodations in place, functional limitations may be reduced or eliminated depending on the context(s) in which they are used.

Conditions that limit function (i.e., impact a functional skill) occur along a continuum. They ebb and flow; we have good days and bad days. Functional limitations are assessed as they occur at their worst. Later in this chapter, we list the functional skill domains typically used in an assistive technology assessment: sensory (visual, auditory, tactile), physical, cognitive, and language. Functional limitations impact our ability to effectively engage in employment-related activities. The general areas include mobility, communication, self-care, self-direction, interpersonal skills, work tolerance, and work skills. Deficits may be defined in terms of areas of sensory, physical, cognitive/psychological, language/communication, as well as the capacity to use assistive technology (Cook & Polgar, 2015).

1. **Mobility:** The capacity for task engagement involving transportation and ambulation within and around the workplace including work-related travel. Common issues encountered include
   a. Disability-related reason for not being able to obtain a driver's license (e.g., seizure disorder, cognitive, or severe motor impairments).
   b. Inability to independently navigate one's common environment or community.
   c. Limitations in the speed and distance in which one can ambulate.
   d. Requires the use of assistive technology to transfer, transport, or ambulate.
2. **Communication:** The capacity for task engagement involving effective expressive or receptive information in the workplace. Common issues encountered include
   a. Requires the use of assistive technology such as augmentative and alternative communication (AAC) technology.
   b. Speech is difficult for others to understand.
   c. Mode of communication is predominantly other than verbal speech (e.g., sign language or AAC device).
3. **Self-care:** The capacity for task engagement involving the performance activities of daily living and the ability to observe safety in work activities. Common issues encountered include
   a. Requires the use of attendant or personal cares provider.
   b. Difficulty with appropriate dress and hygiene (e.g., getting dressed, selecting weather appropriate clothing).
   c. Requires a strict medication or dietary regimen that must be observed during work.
4. **Self-direction:** The capacity for task engagement involving making adaptive and effective decisions in the workplace. Common issues encountered include
   a. There is a court appointed legal guardian and/or representative payee who facilitate effective decision-making.
   b. A lack of disability insight and how this impacts cognition and behavior as it relates to employment.
   c. Requires frequent and/or routine reminders, prompts, and cues for on-task behavior.
   d. Difficulty comprehension or relies heavily on others' regarding their Great 8.
5. **Interpersonal skills:** The capacity for task engagement involving the development and maintenance of interpersonal relationships in the workplace (National Consortium of Interpreter Education Centers, 2017). Common issues encountered include

   a. History of interpersonal conflict (e.g., difficulties with authority figures, can't get along effectively with coworkers).
   b. Exhibits behavior which might be difficult for others to understand (e.g., Tourette's syndrome, motor dysfunction, traumatic brain injury sequelae).
   c. Difficulty connecting with others in a meaningful way.
   d. Significant fear and avoidance around social interaction.
6. **Work tolerance:** The capacity for task engagement involving the performance of the physical demands required of a client in a job [e.g., stamina, gross/fine motor skills, somatosensory, range of motion (ROM), and muscle strength, tone, reflex] (National Consortium of Interpreter Education Centers, 2017). Common issues encountered include
   a. Difficulty with balance and coordination.
   b. Prone to pressure sores when seated without proper seating/positioning.
   c. Due to disability the client needs a job that is predominantly sedentary, allows for frequent rest periods, a flexible work schedule, and/or frequent position changes.
   d. Disability causes pain, stiffness, weakness, and/or swelling which must be accommodated.
7. **Work skills/history:** The capacity for task engagement involving entry-level work. In other words, there is a reasonable expectation that an individual can engage in work activities without training (National Consortium of Interpreter Education Centers, 2017). Common issues encountered include
   a. A history of problems with task initiation, maintenance, completion, organization, time management, accuracy, and production rate (due to disability-related factors only).
   b. Requires frequent and/or routine workplace accommodations not typically made for most workers.
   c. Work history demonstrates frequent negative elements as they relate to work (e.g., many short-term work experiences, lack of good references/poor reputation, long-term periods of unemployment).

When considering job accommodations to address functional limitation for condition such as clinical depression, it would be important to note both extremes of your personal experience with depression. Those times on the higher functioning end of the continuum when you may not have wanted to socialize with others (but did so anyway) as well as the times when it was so difficult you could not get out of bed, wept incessantly and for no apparent reason, avoided going out in public, and would not eat anything for days. This is critical to determining the severity of a functional limitation. Without this information, it can be difficult, if not impossible, to determine effective interventions toward goal attainment and successful vocational outcomes.

ASSISTIVE TECHNOLOGY FUNCTIONAL SKILL DOMAINS
SENSORY SKILLS DOMAIN: VISUAL FUNCTIONAL CAPACITY

- Visual field deficit: Experienced in one of two ways: loss of peripheral or central vision.
  - Peripheral vision loss: Narrowing of the visual field. Typically, age-related.
  - Central vision loss: Loss resulting in the inability to see what an individual is looking at directly.
- Visual acuity: The clarity with which an individual can see objects in their environment.
  - Myopia (near-sightedness): Inability to focus on distant objects.
  - Hyperopia (far-sightedness): Inability to focus on near objects.
  - Presbyopia: Inability to focus on near objects (age-related visual change).
- Visual tracking: The ability to track a moving object.
  - Coordination of the eyes: Capacity of both eyes to work together. Includes fusing images from both eyes into a single three-dimensional image and proper eye muscle control/alignment of the eyes.
  - Vertical/horizontal tracking: Tracking on both vertical and horizontal planes.
  - Smoothness of movements: The fluidity of the tracking movement.
  - Tracking initiation delay: Presence of any interstitial pause in tracking movement initialization.
  - Tracking without head movement: Visual tracking without head movement.
  - Visual scanning: In visual scanning, the ability to image the environment to collect data; an object being viewed does not move, rather, the eyes move.
- Visual contrast: Differentiation of an object from its background (including information appearing on a visual display or monitor).
  - Contrast augmentation: Contrast enhancement tends to be a greater need with age.
- Visual accommodation: Capacity of the eyes to refocus on objects such as from near-objects to far-objects.
  - Eye muscle coordination: This is necessary to achieve effective vision.
- Visual perception: The assignment of meaning to visual stimuli including skills such as spatial associations, depth perception, figure–ground discrimination, and form appreciation/constancy.
  - Figure–ground perception: Discrimination between foreground and background.
  - Spatial relations: Comprehension of objects as they relate to each other.
  - Form constancy (recognition): Comprehension of the constancy of an object regardless of whether the object position or viewer perspective is changed or changing.

- Color vision: The capacity to differentiate objects based on the reflection of light waves (colors)—Light is detected by rod and cone cells of the retina. Rod cells are responsible for low light vision and cone cells are responsible for color and detail vision.
  - Scotopic vision: Vision in very low light conditions—light is detected by rod cells and are maximally sensitive at 500 nm.
  - Photopic vision: Vision in brighter light conditions—Light is detected by cone cells maximally sensitive at 550 nm.
  - Mesopic vision: Combination vision—rods and cones—shifts in color perception from low to brighter light.

SENSORY SKILLS DOMAIN: AUDITORY FUNCTIONAL CAPACITY

- Auditory thresholds: Auditory thresholds are measured as a product of both frequency and amplitude.
  - Frequency/pitch/hertz (Hz): Measurement of sound waves: cycles per second.
        In humans frequency range is 20—20,000 Hz.
  - Amplitude/intensity/loudness:
    - Measurement of sound waves: decibels (dB)—sound power/loudness
    - 0 dB = best hearing 50% of the time.
    - 120 dB = painfully intense for most people with normal hearing.
      - Normal hearing: 0−20 dB
      - Mild hearing impairment: 20−40 dB
      - Moderate hearing impairment: 40−55 dB
      - Moderately severe hearing impairment: 55−70 dB
      - Severe hearing impairment: 70−90 dB
      - Profound hearing impairment: >90 dB
  - Auditory perception: Similar to visual figure−ground perception, auditory perception is the capacity to differentiate sound from background or ambient sound.

SENSORY SKILLS DOMAIN: TACTILE (SOMATOSENSORY) FUNCTIONAL CAPACITY

- Tactile function: Touch sensation through either active or passive touch.
  - Active touch: Engaging an object or person through touch.
  - Passive touch: Being the passive receiver of a touch from an object or person.
- One-two point discrimination: The simultaneous application of pressure from two points.
  - The ability to differentiate between two points of stimuli to a body part.

- Touch perception: The ability to detect touch and differentiate between light to deep pressure.
  - Light versus deep pressure: Capacity to detect active or passively received touch while differentiating light or deep pressure.
- Temperature: The ability to detect hot or cold sensations.
- Pain perception: The ability to perceive pain tends to be sharp or dull in nature.
  - Sharp pain: Light pressure pain sensation (needle puncture).
  - Dull pain: Deep pressure pain sensation (blunt impact).
- Joint position (Proprioception): Ability to determine the position of a joint or limb in relation to the body.
- Localization of tactile stimulation: The determination of a specific site or location on the body of a tactile sensation.

PHYSICAL SKILLS DOMAIN

- Range of motion (ROM): Distance (linear or angular) a joint travels in relation to the body.
  - Passive ROM: Movement of a joint without active assistance. Muscles are not actively engaged in the motion.
  - Active-assistive ROM: Assistance in moving a joint—can be a person or mechanical assistance. Can be used to avoid further injury.
  - Active ROM: Use of musculature to move a joint—little to no protection is needed (e.g., strengthening exercises).
- Muscle strength: The capacity for a muscle to contract to produce force.
  - Measured in grades on a five-point scale based on resistance.
    - 0/5—no obvious muscle contraction.
    - 1/5—obvious contraction without movement.
    - 2/5—can contract without or with reduced gravity.
    - 3/5—fully contract muscle and move against gravity throughout full ROM.
    - 4/5—muscle yields to maximum resistance-unable to maintain contraction against maximum resistance.
    - 5/5—muscle function is normal and maintains contraction against full resistance.
- Muscle tone: The degree of tension present in a muscle. Muscle tone provides the foundation by which bodily movements occur and posture is maintained opposite the force of gravity.
  - Hypertonia: Condition of abnormally increased muscle tone—hyper excitability of the stretch reflex.
  - Hypotonia: Condition of low muscle tone—low tension present in the muscle, low resistance to muscle stretching.
- Obligatory movements (presence of): Key reflexes toward postural, positional, and alignment control including balance and degree of support required to maintain them.

- Dependent: Full support required due to lack of key reflexes.
  - Assisted: Able to partially support due to some key reflexes.
  - Independent: Requires no assistance due to fully responsive key reflexes.
- Motor skills and planning: Motor function involving the organization and planning of complex muscle movements.
  - Gross motor function: Use of large muscle groups for task performance.
  - Fine motor function: Use of small muscle groups for task performance.
  - Complex motor planning: Combination of executive functioning and complex motor movement including learning from feedback and error correction.

COGNITIVE SKILLS DOMAIN

- Perception: Relates to how environmental stimuli are perceived as received from the senses.
- Orientation: Cognitive status evaluation regarding an individual's relative cognitive disposition toward person, place, time, and quantity.
  - Person: Being cognizant of self and others in context.
  - Place: Being cognizant of context or environment (knowing one's immediate geographical location in relation to an expanded geographical scope).
  - Time: Being cognizant of not only the current date and time but also consciousness regarding time management. This includes the capacity to utilize chronological sequencing, time allocation, and recognition of the relative passage of time.
  - Quantity: Being cognizant of numerical increments, calculations, and sequencing; use of and comprehension of numbers.
- Attention: The process of discrimination among environmental stimuli to determine the focus of cognitive resources.
  - Signal detection: The capacity to discriminate the presence of discrete stimuli.
  - Vigilant attention: The capacity to sustain focus on stimuli.
  - Searching: Capacity to discriminate the potential appearance of a discrete stimulus among a field of stimuli (scanning).
  - Selective attention: The capacity to identify a discrete stimulus and track it among a field of stimuli.
  - Divided attention: Allocating cognitive resources toward attending to more than one stimulus.
- Memory: Recall of past learned or experienced information.
  - Encoding: Process of actively attending to information toward storage for later retrieval.
  - Storage: Process of consolidation of information for future retrieval.
  - Retrieval: Process of consciously attending to information pulled from memory.

- Sensory memory: Shortest term memory with the least degree of memory storage.
- Short-term memory: Moderate term memory and capacity for memory storage.
- Long-term memory: Longer term memory with a high capacity for memory storage and longevity for retrieval.
- Implicit memory: Memory which augments task performance due to practice effects and which derive without conscious awareness.
- Explicit memory: Memory which derives from conscious retrieval.
- Recall: Retrieval of stored information into awareness without the benefit of cuing or prompting.
- Recognition: Retrieval of stored information into awareness with the benefit of cuing or prompting.
- Problem solving: The primary function of problem solving is to delineate the means by which a problem is addressed and a solution identified.
  - Problem identification: The process of becoming aware of and delineating a problem.
  - Judgment: The formulation of an opinion following deliberation.
  - Decision making: The process of determining a course of action from a number of alternative or competing courses of action.
  - Deductive reasoning: A form of logical reasoning moving from a more broad-based reasoning to the more specific. Often referred to "top-down" logic.
  - Inductive reasoning: A form of logical reasoning moving from the more specific or detail-oriented to a more broad-based and general type of reasoning. Often referred to "bottom-up" logic.
  - Planning: The process of developing goals, policies, and procedures toward a defined outcome (may involve task foci of initiation, maintenance, completion, organization, accuracy, production, and safety).
  - Evaluation: The systematic process of making a judgment or assessment regarding the significance and value of something.
  - Knowledge transfer: The process by which expertise, knowledge, skills, or capabilities are transferred to others (i.e., shared with them).
- Knowledge representation: The process by which knowledge and semantic information is encoded, how external information is represented cognitively.
  - Declarative knowledge: Knowledge about something. Typically defined as knowing the facts or the "what" regarding something.
  - Procedural knowledge: Knowledge regarding how to do a task or activity. Procedural knowledge is descriptive in nature regarding the "how to" of something.
  - Categorizing: The process of organizing or classifying things into groups, classes, and categories based on common characteristics or discrete features which differentiate the members of one assortment from another. Focused on the relationship between things.

- Sorting: The systematic arrangement of things into groups or types based on predefined characteristics. Focus is on organizing things based on defined characteristics rather than on any relationship between categories.
  - Sequencing: Arrangement of things based on an order or in a series. Focus is on sequencing things based on a set of predetermined rules.
- Executive function: Set of higher order processes for cognitive regulation and control including analysis, problem-solving, planning, organizing, and monitoring functions.
- Language: The capacities to systematically express, receive, and organize language elements and abstract concepts, and construe meaning from these.

LANGUAGE SKILLS DOMAIN

- Expressive: Verbal, nonverbal and written language including sentence structure (does not include speech production).
- Receptive: Includes symbol/word recognition and comprehension of simple instructions.
- Sequencing (rational information processing): Ability to order correctly spoken or written words, numbers, or symbols.
- Symbolic representation: A process whereby symbols denote abstract concepts and express meaning—also, label-referent pairing (encoding).
- Language element combination: Ability to combine componential parts of language to produce a natural flow (of expressive language).
- Usage of codes (AT-related): Ability to use codes as representations of ideas and objects for communication.
- Categorization: Ability to select, differentiate, and organize task-relevant stimulus features.
- Matching: Recognition of different types of language elements and making associations between conceptually similar elements (e.g., image of stairs = the word stairs = a symbol for stairs).
- Level of social interaction: The degree to which one communicates with others—this is closely associated with level of learning and is critical to social development also.
- Motor speech: Coordination and sequencing of neuromuscular and motor control resulting in the production of speech (includes respiratory, phonatory, resonatory, and articulatory systems).
- Pragmatic language: The competence with which language is used in context, infer meaning from others' language, and chose syntax.
- Syntax: Capacity to structure and organize the elements of language.
- Semantics: Capacity to understand and use the meaning in language—meaning versus structure of language.

The Assistive Technology Functional Skill Domains above are adapted from Cook, A. M., & Polgar, J.M. (2015). *Assistive technologies: Principles and practice* (4th Edition). St. Louis, MO: Mosby. Used with Permission.

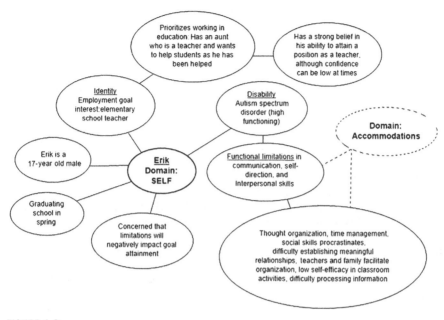

FIGURE 1.3

Erik case study map: self.

There is no "cookie cutter" approach to determining functional limitations. They drive the employment planning process relative to our personal circumstances, the nature of the disability, the employment goal, employment contexts, task engagement, accommodations considerations, and how well we navigate these complexities. Fig. 1.3 illustrates a case study map for Erik regarding considerations inherent within the self.

SUMMARY

The self represents the innate qualities an individual brings to employment. As one of the five domains comprising the Accommodation System (AS), it informs service provision regarding those qualities inherent and closely associated with the self. The AS is dynamic and each of the domains (i.e., self, others, environment, accommodations, and situated experience) interact with and influence the others. Conceptualizing the self helps facilitate employment-related service provision (including accommodations like assistive technology). Effective service provision invariably begins with understanding an individual and their unique needs as they begin the process of looking for or maintaining employment following disability (habilitation or rehabilitation). The AS sets the stage for the provision of accommodation services within the context of the employment of people with disabilities.

REFERENCES

Adya, M., Samant, D., Scherer, M. J., Killeen, M., & Morris, M. W. (2012). Assistive/rehabilitation technology, disability, and service delivery models. *Cognitive Processing*, *13* (1), 75−78.

Ajzen, I. (1991). The theory of planned behavior. *Organizational Behavior and Human Decision Processes*, *50*(2), 179−211.

American Psychiatric Association. (2013). *Diagnostic and statistical manual of mental disorders* (5th ed). Washington, DC: Author.

Bandura, A. (1971). *Social learning theory* (pp. 1−46). Morristown, NJ: General Learning Press.

Bandura, A. (1986). *Social foundations of thought and action: A social cognitive theory*. Englewood Cliffs, NJ: Prentice-Hall.

Bandura, A. (1991). Self-efficacy mechanism in physiological activation and health-promoting behavior. In J. Madden, IV (Ed.), *Neurobiology of learning, emotion and affect* (pp. 229−270). New York, NY: Raven.

Bandura, A. (1997). *Self-efficacy: The exercise of control*. New York, NY: W. H. Freeman and Company.

Bandura, A. (2001). Social cognitive theory: An agentic perspective. *Annual Review of Psychology*, *52*, 1−26.

Bond, G. R. (2004). Supported employment: Evidence for an evidence-based practice. *Psychiatric Rehabilitation Journal*, *27*(4), 345−359.

Bridges, D. (1993). Transferable skills: A philosophical perspective. *Studies in Higher Education*, *18*(1), 43.

Bruner, J. S. (1966). *Toward a theory of instruction*. Cambridge, MA: Harvard University Press.

Carr, H. A., & Kingsbury, F. A. (1938). The concept of ability. *Psychological Review*, *45* (5), 354−376.

Chan, F., Sasson Gelman, J., Ditchman, N., Kim, J. H., & Chiu, C. Y. (2009). The world health organization ICF model as a conceptual framework of disability. In F. Chan, E. Da Silva Cardosa, & J. A. Chronister (Eds.), *Understanding psychosocial adjustment to chronic illness and disability* (pp. 23−50). New York, NY: Springer Publishing Company.

Cobigo, V., Lachapelle, Y., & Morin, D. (2010). Choice-making in vocational activities planning: Recommendations from job coaches. *Journal of Policy and Practice in Intellectual Disabilities*, *7*(4), 245−249.

Cook, A. M., & Polgar, J. M. (2015). *Assistive technologies: Principles and practice* (4th Edition). St. Louis, MO: Mosby.

Csikszentmihalyi, M. (1975). *Beyond boredom and anxiety: Experiencing flow in work and play*. San Francisco, CA: Jossey-Bass Inc.

Csikszentmihalyi, M., Abuhamdeh, S., & Nakamura, J. (2005). Flow. In A. J. Elliot, & C. S. Dweck (Eds.), *Handbook of competence and motivation* (pp. 598−608). New York, NY: The Guilford Press.

Csikszentmihalyi, M., & Csikszentmihalyi, I. S. (1988). Introduction to part III. In M. Csikszentmihalyi, & I. S. Csikszentmihalyi (Eds.), *Optimal experience: Psychological studies of flow in consciousness* (pp. 183−192). New York, NY: Cambridge University Press.

Dawis, R. (1996). Vocational psychology, vocational adjustment, and the workforce: Some familiar and unanticipated consequences. *Psychology, Public Policy, and Law*, 2(2), 229−248.

Dawis, R. V., & Lofquist, L. H. (1984). *Psychological theory of work adjustment: An individual-differences model and its applications*. Minneapolis, MN: University of Minnesota Press.

Day, H., & Jutai, J. (1996). Measuring the psychosocial impact of assistive devices: The PIADS. *Canadian Journal of Rehabilitation*, 9, 159−168.

Deci, E. L., & Moller, A. C. (2005). The concept of competence. In A. J. Elliot, & C. S. Dweck (Eds.), *Handbook of competence and motivation* (pp. 579−597). New York, NY: The Guilford Press.

Dewey, J. (1963). *Reconstruction in philosophy*. Boston, MA: The Beacon Press.

Erikson, E. (1980). *Identity and the life cycle*. New York, NY: W. W. Norton & Company.

Erikson, E. (1997). *The life cycle completed*. New York, NY: W. W. Norton & Company.

Federici, S., Meloni, F., & Corradi, F. (2012). Measuring intellectual functioning. In S. Federici, & M. J. Scherer (Eds.), *Assistive technology assessment handbook* (pp. 25−48). Boca Raton, FL: CRC Press.

Fenton-O'Creevy, M., Dimitriadis, Y., & Scobie, G. (2015). Failure and resistance at boundaries: The emotional process of identity work. In E. Wenger-Trayner, M. Fenton-O'Creevy, S. Hutchinson, C. Kubiak, & B. Wenger-Trayner (Eds.), *Learning in landscapes of practice: Boundaries, identity, and knowledgeability in practice-based learning* (pp. 33−42). New York, NY: Routledge.

Feuerstein, R., Falik, L. H., & Feuerstein, R. S. (2015). *Changing minds & brains: The legacy of Reuven Feuerstein*. New York, NY: Teacher's College Press.

Feuerstein, R., Feuerstein, R. S., & Falik, L. H. (2010). *Beyond smarter: Mediated learning and the brain's capacity for change*. New York, NY: Teacher's College Press.

Fried, L. P., Ferrucci, L., Darer, J., Williamson, J. D., & Anderson, G. (2004). Untangling the concepts of disability, frailty, and comorbidity: Implications for improved targeting and care. *The Journals of Gerontology Series A: Biological Sciences and Medical Sciences*, 59(3), 255−263.

Giesen, J. M., & Hierholzer, A. (2016). Vocational rehabilitation services and employment for SSDI beneficiaries with visual impairments. *Journal of Vocational Rehabilitation*, 44(2), 175−189.

Gollwitzer, P. M., & Kirchhof, O. (1998). The willful pursuit of identity. In J. Heckhausen, & C. Dweck (Eds.), *Motivation and self-regulation across the lifespan* (pp. 389−423). New York, NY: Cambridge University Press.

Jette, A. M. (2006). Toward a common language for function, disability, and health. *Physical Therapy*, 86(5), 726−734.

Josselson, R. (1992). *The space between us: Exploring the dimensions of human relatedness*. San Francisco, CA: Jossey-Bass Inc. Publishers.

Kanfer, R., & Ackerman, P. L. (2005). Work competence: A person-oriented perspective. In A. J. Elliot, & C. S. Dweck (Eds.), *Handbook of competence and motivation* (pp. 336−353). New York, NY: The Guilford Press.

Kasl, E., & Elias, D. (2000). Creating new habits of mind in small groups. In J. Mezirow (Ed.), *Learning as transformation* (pp. 229−252). San Francisco, CA: Jossey-Bass.

Kasnitz, D., Switzer Fellow, M., & Shuttleworth, R. P. (2001). Introduction: Anthropology in disability studies. *Disability Studies Quarterly*, 21(3), Summer, 2−17.

Kingsbury, F. A., & Carr, H. A. (1939). The concept of directional disposition. *Psychological Review*, *46*(3), 199–225.

Kosciulek, J. F. (2007). A test of the theory of informed consumer choice in vocational rehabilitation. *Journal of Rehabilitation*, *71*(2), 41–49.

Lange, M. (2008). Clinical fundamentals. In M. Lange (Ed.), *Fundamentals in assistive technology: An introduction to assistive technology implementation in the lives of people with disabilities (4th ed.*, pp. 1–31). Arlington, VA: RESNA Press.

Langer, E. J., & Dweck, C. S. (1973). *Personal politics: The psychology of making it.* Englewood Cliffs, NJ: Prentice-Hall, Inc.

Lave, J., & Wenger, E. (1991). *Situated learning: Legitimate peripheral participation.* Cambridge, UK: Cambridge University Press.

Leslie, A. M. (1995). A theory of agency. In A. J. Premack, D. Premack, & D. Sperber (Eds.), *Causal cognition: A multidisciplinary debate* (pp. 131–149). Oxford: Clarendon Press.

Lewin, K. (1999). Level of aspiration. In M. Gold (Ed.), *The complete social scientist* (pp. 137–182). Washington, DC: American Psychological Association.

Littman-Ovadia, H., Lazar-Butbul, V., & Benjamin, B. A. (2014). Strengths-based career counseling: Overview and initial evaluation. *Journal of Career Assessment, 22,* 403–419.

Meyer, A., Rose, D. H., & Gordon, D. (2014). *Universal design for learning: Theory and practice.* Wakefield, MA: CAST Professional Publishing.

Minor, C. A., & Bates, P. E. (1997). Person-centered transition planning. *Teaching Exceptional Children*, *30*(1), 66–69.

National Consortium of Interpreter Education Centers. (2017, July). *Vocational rehabilitation glossary.* Retrieved from: ⟨http://www.interpretereducation.org/vr_tk/⟩.

National Research Council. (2000). *How people learn: Brain, mind, experience, and school: Expanded edition.* Washington, DC: National Academies Press.

Peterson, C. (2006). The values in action (VIA) classification of strengths. In M. Csikszentmihalyi, & I. S. Csikszentmihalyi (Eds.), *A life worth living: Contributions to positive psychology* (pp. 29–48). New York, NY: Oxford University Press.

Peterson, M. (1995). Ongoing employment supports for persons with disabilities: An exploratory study. *Journal of Rehabilitation*, *61*, 58.

Prince, M. J. (2017). Persons with invisible disabilities and workplace accommodation: Findings from a scoping literature review. *Journal of Vocational Rehabilitation*, *46*(1), 75–86.

Sagiv, L., & Schwartz, S. H. (2000). Value priorities and subjective well-being: Direct relations and congruity effects. *European Journal of Social Psychology*, *30*(2), 177–198.

Shay, A. F., Anderson, C. A., & Matthews, P. (2017). Empowering youth self-definition and identity through assistive technology assessment. *Vocational Evaluation and Work Adjustment Association Journal*, *41*(2), 78–88.

Sylvia, P. J. (2008). Interest—the curious emotion. *Current Directions in Psychological Science*, *17*(1), 57–60.

Tilson, G., & Simonsen, M. (2013). The personnel factor: Exploring the personal attributes of highly successful employment specialists who work with transition-age youth. *Journal of Vocational Rehabilitation*, *38*(2), 125–137.

Vocational Rehabilitation. 80 (181) Fed. Reg. § 361.1 (August 19, 2016) (to be codified at 34 CFR Pts 361, 363, and 397).

Wicklund, R. A., & Gollwitzer, P. M. (1982). *Symbolic self-completion*. New York, NY: Routledge.

Wolfe, R. (2001). The experience sampling method and career counseling: The interrelations of situated experience, work values, and career orientation among adolescents. *Unpublished doctoral dissertation*. Chicago, IL: University of Chicago.

Wolff-Branigin, M., Daeschlein, M., Cardinal, B., & Twiss, M. (2000). Differing priorities of counselors and customers to a consumer choice model in rehabilitation. *Journal of Rehabilitation*, *66*(1), 18−22.

World Health Organization. (2001). *International classification of functioning, disability, and health*. Geneva: Author.

Accommodation System: Other

Anthony Shay

Capacity Building Specialist, Assistive Technologist, and Rehabilitation Specialist, University of Wisconsin-Stout Vocational Rehabilitation Institute (SVRI), Menomonie, WI, United States

THE SOCIAL DOMAIN

Whether we are engaged in our first employment experience having always had a disability or reentering the world of work after having acquired one we must navigate the social dimension while at work. Social engagement is the cornerstone of vocational competence and job maintenance. Developing a sense of mutuality and connectedness with others requires of us to recognize ourselves in the spaces where work occurs and to be aware of others' perspectives as they relate to the same spaces and work activities. In short, we must maintain an awareness of self and others. We construct meaning in the workplace through shared experiences with the people with whom we interact, the activities in which we are engaged, and the environments within which all this takes place. The accommodation system (AS) views interpersonal activity and the environments in which they occur as distinct from each other. Understanding others as selves in their own right and differentiating contextual artifacts from the people who create and perpetuate them serves to more effectively conceptualize the content and process of effective service delivery.

THE SELF IN RELATION TO OTHERS

Human interaction forms the basis by which reality is mutually created. Relationships "are the ground of our lives and, like the air that we breathe, we absorb and act on them" (Josselson, 2007, p. 29). Regardless of the context we find ourselves in, it is the social dimension of practice and identity building that enables the shaping and realization of competence (Wenger-Trayner & Wenger-Trayner, 2015) and reinforces resiliency (Bandura, 1997). Bandura (1997, p. 6) finds, "human adaptation and change are rooted in social systems." The quality and impact of relationships on our lives are always changing and may go unnoticed. The social dynamic within the workplace (as with all social interaction) is based on this human interconnectedness (Josselson, 2003).

Assistive Technology Service Delivery. DOI: https://doi.org/10.1016/B978-0-12-812979-1.00002-3

We can consider interpersonal connectedness along eight rational dimensions. These elements reflect a person's relational developmental needs along a continuum based on the absence or excess of the dimensions. These are broken down into two sets of four dimensions. The first four are present very early in life with the second set developing as we mature: holding, attachment, passionate experience, eye-to-eye validation; idealization/identification, embeddedness, mutuality/resonance, and tending/care. These dimensions represent movement in developmental awareness away from an exclusive focus on the self to an increasing awareness of the perspectives of others:

- Holding: The fundamental experience of safety, support, and trust that needs will be met;
- Attachment: A discriminatory focus of holding which responds in kind to our attention;
- Passionate experience: A more intense expression of interpersonal connection characterized by emotional arousal;
- Eye-to-eye validation: An affirmation of self through social validation;
- Idealization/identification: The internal process of building connectedness while recognizing others' differences in relation to the self;
- Embeddedness: Alignment, visioning, and engagement within a social group;
- Mutuality/resonance: A complete, growth oriented, and mutual sense of companionship or bond with others characterized by a growing sense of security, openness, and intimacy;
- Tending/care: Connectedness stemming from an expression of the self in giving while attending to the needs of the other (Josselson, 2003).

Our connections with others on the job serves to both build relationships and facilitate task engagement (Drucker, 2010). Through mutual exchanges, we are cocreators of our vocational realities that provide resources for and constraints upon our actions (Bandura, 1997; Wicklund & Gollwitzer 1982). We seek both to be connected to others and to maintain our individuality. This sets the stage for a social dialectic which can be a facilitator or detractor of motivation. The more chaotic and controlling our social networks become, the greater the negative impact is on our motivation, identity development, and meaningful relationship building (Deci, 1995). We ascribe meaning to our behavior and that of others even as they ascribe meaning to our own. Changing perspectives can color how we perceive ourselves and others. How well we navigate social interaction—our interpersonal orientation—has an impact on true integration in community-based employment.

Selman (2003) defines "interpersonal orientation," as a reconciliation of self versus other. It is an ongoing inner dialog based on personal experience. Selman suggests that this is how we accommodate someone's wishes or how we assert our own will. He defines this as a *self or other* transformational interpersonal *orientation*. Furthermore, this assertion of will is an interpersonal *negotiation* when self and other are considered together. As these orientations develop, they "become both well differentiated and integrated" (Selman, 2003, p. 37). A well-developed sense of collaboration (i.e., optimal social integration) necessarily involves others and cannot be achieved without them. The goal is to balance our need for autonomy with our need for relatedness. When we are able to balance these needs, we will have developed a sense of social competence (Selman, 2003, p. 37). Learning is inherently a social process (Vygotsky, 1978, p. 88). As our social milieus change so to must we adapt and develop to meet the changes. Through our vocational interactions, we increase the potential for growth in both ourselves and in the others.

THE DIALOGICAL OTHER

The dialogical other is a reminder of our interconnectedness with Others. Everyone we encounter in the workplace is a self in their own right. We must contend with our inner voice and the voices of others populating our work world (i.e., what we tell ourselves and what we hear from others). We must also contend with the inner voices of the other selves who populate the inner world of our mind (i.e., what we think other people are saying). Learning occurs through authentic connectedness with others. Communication (i.e., dialog) forms the basis for learning but requires of us reflection and action for social awareness to develop. Language and communication are the overt artifacts of our direct connectedness with those around us. With these, we build relationships; the implicit sense of "being" with another person that resides beneath the overt. These artifacts cannot simply be reduced to an act of "depositing" ideas from one person into another nor can it simply be an exchange of ideas to be "consumed" by those with whom we communicate (Freire, 2000, pp. 88−89). Language (verbal and nonverbal) is an active engagement with those around us through which we lay the groundwork for building relationships— the unspoken product of an active ongoing mutual construction of our sense of "self" in relation to the "other" (Josselson, 2007, p. 4).

OTHERS IN RELATION TO THE SELF

We may also seek to understand the other in relation to the self rather than trying to see things from their point of view alone. In other words, the self seeks to understand the world from a shared perspective. This form of connectedness moves us "from thinking in terms of 'I' to thinking in terms of 'we'." It brings the self and the other closer together (Selman, 2003, p. 28). According to Erik Erikson, "each person is a center of awareness in a universe of communicable experience." Thinking in terms of "we" requires of us to share a context with others and to communicate effectively with them (Erikson, 1998, p. 87). This has a humanizing effect toward building inclusive experiences. Although we seek inclusion, we must maintain our sense of self. The differentiation of self and other allows the revision of the "inner model of ourselves in light of our interpersonal experiences... [to] realize ourselves only in, through, and with others" (Josselson, 1992, p. 19). This strengthens us emotionally and reinforces resiliency allowing us to test our boundaries and take interpersonal risks (Josselson, 1992).

Others exert social influence toward adherence to group norms. We assign meaning to our belief system and those of others by measuring them against our own (Lewin, 1997). Relationships of an unofficial or informal nature within the workplace may have the effect of reinforcing dysfunctional or ineffective workplace interpersonal practices (Juniper, 1995; Lave & Wenger, 1991, p. 64). These less than ideal circumstances can lead to seemingly insurmountable impediments to employment. The Johari Window provides a frame through which we gain perspective on this negotiation.

THE JOHARI WINDOW

The complexities of our interaction with others in the workplace can be difficult to describe. We cannot overlook their importance. Job maintenance depends on successful navigation of the social domain. Erik Erikson described the importance of understanding interpersonal relationships and their impact on what he referred to as the self-observing *I* and the shared *we* (Erikson, 1998, p. 87). The Johari Window, developed by Joe Luft and Harry Ingham (from whom the model derives the name *Johari*), is well suited to a discussion of human interaction, communication, and interpersonal awareness (Luft, 1969). We present it here as a useful means of describing the interpersonal nature of integrating the self with others in a work environment. Although the model is a simplification of the complex nature of human interaction, it does provide a way to frame the discussion of interpersonal disclosure irrespective of an individual's culture or background (Hanson, 1973).

The Johari Window is a model for soliciting and providing feedback about the self to others. It consists of four window panes (i.e., quadrants) analogous to how we look out into and view the world. The window is divided into rows and columns each consisting of two quadrants. The rows represent aspects of the self, and the columns represent aspects of the other. The four quadrants where self and other intersect (the window panes) represent aspects of the self's subjective experience relative to interaction and awareness (Hanson, 1973; Luft, 1969; Schein, 2013, pp. 103−109). These include the open, blind, hidden, and unknown areas (UAs) (see Fig. 2.1).

The open area (OA, upper left): Represents those things I know about myself and things others know about me. Overt communication may be free and open between the self and others. This objective information (e.g., verbal, observable) is public and available to anyone (Hanson 1973; Luft, 1969).

The blind area (BA, upper right): Represents those things I do not know about myself and things others know about me. We may be unaware of the information we communicate to others. This may take the form of verbal or nonverbal

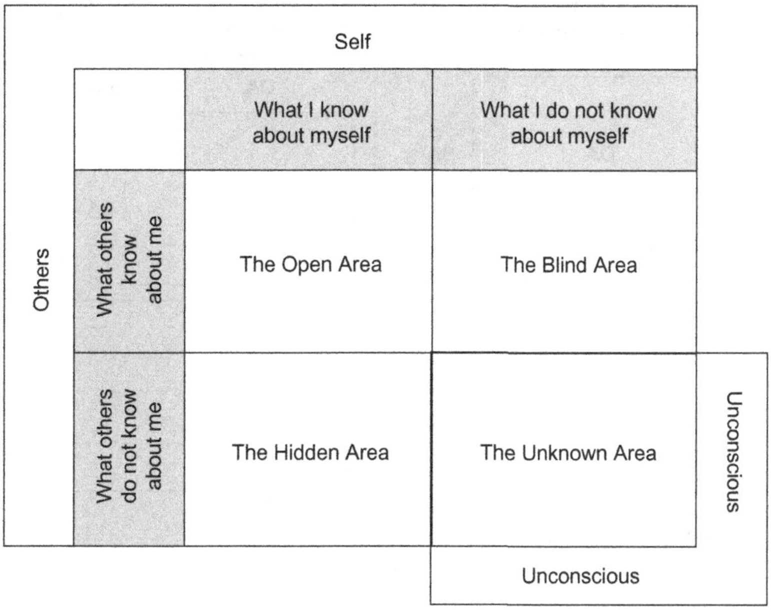

FIGURE 2.1

The Johari Window.

cues, mannerisms or behavior, conversational style, or our individual manner of relating to others (Hanson 1973; Luft, 1969).

The hidden area (HA, lower left): Represents those things I know about myself and things others do not know about me. At times, we desire to keep information hidden from others. This may be due to a fear of rejection, retaliation, or hurt if information is shared with others (e.g., fear of being judged negatively by others due to personal convictions, attitudes, feelings, opinions, stigma, etc.) (Hanson 1973; Luft, 1969).

The Unknown Area (UA, right): Represents those things I do not know about myself and things others do not know about me. There may also be information that will never be known completely by the self or others. Through the exchange of feedback between self and others, information is shared, insight is gained, and the UA diminishes. As this exchange continues, the OA becomes larger. Our unknown skills and abilities can reside in the UA. It is possible to have a Johari Window profile in which there would be no unknown information, although this is extremely unlikely. As a result, the model depicts the unknown window with extensions illustrating information which may never be fully known to the self. In a Freudian sense, this domain extension is known as the **unconscious area** (Hanson 1973; Luft, 1969).

An effective way of integrating with others in social settings is to increase the size of the OA while reducing the size of the other windows. However, when it

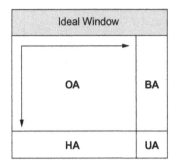
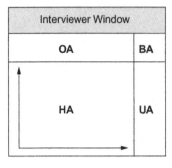
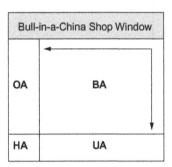
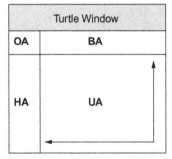

FIGURE 2.2

Johari Window with markedly differing styles of self-other awareness.

comes to soliciting or providing feedback, we tend to do one more than the other. The size of the OA, then, is a function of how much feedback is solicited by others and given by the self. Fig. 2.2 illustrates the four windows demonstrating markedly differing styles of self-other awareness. These windows are labeled as the ideal window (large OA), interviewer window (large HA), the bull-in-a-china-shop window (large BA), and the turtle window (large UA) (Hanson, 1973).

The ideal profile: An ideal profile is characterized by a large degree of trust between self and other. The OA increases as the amount of trust increases. An ideal profile affords less opportunity for others to misinterpret behavior. With clear and open communication others spend less time engaged in guesswork and more time in meaningful interaction (Hanson, 1973). This degree of openness may be threatening for some. It also tends to reduce the need for superficial and insincere behavior, important considerations in the development of meaningful relationships.

The interviewer profile: Interviewers are good information seekers. They are not, however, good at offering any in return. What they do offer others is equal to the social norm. They seek to maintain whatever others will accept as a reasonable level of participation. The size of their HA is inversely related to the amount of feedback flowing out from the individual (Hanson, 1973). It is difficult to know where these individuals stand on issues. When we encounter others like this, we may ourselves react to them with distrust and irritation. We may also find ourselves wanting to withhold information from them since this is the interviewer's stock-in-trade.

The bull-in-china-shop profile: This type of person maintains a level of interpersonal interaction primarily through providing feedback to others. However, they solicit very little from others in return. They freely offer others their stance and feelings on the issues. The bull-in-china-shop is only too eager to share with us those qualities of the self which we are unaware of or not ready to accept (Hanson, 1973). They may lash out at others or be critical of them. They may appear to be insensitive to others' feedback or may appear too disinterested in what others have to say. The bull-in-china-shop tends to respond to feedback in such a way as to make others reluctant to participate in social activities or to provide accurate, if any, feedback.

The turtle profile: Turtles tend to know little about themselves. They also prefer that others know little about them. These are the silent members of a work group. They observe rather than participate. Others learn where turtles stand on issues or how they feel more generally only after much effort. They offer little in the way of sentiment. Establishing, affirming, and maintaining meaningful relationships is exceedingly difficult. A defining characteristic of a turtle is their insular nature. They tend to seek or reinforce marginalization making relationship building not only difficult but also uncomfortable for others (Hanson, 1973).

Silence should never be used as the sole criterion for identifying a turtle. Because working in groups can be threatening to some people, we should take care to be sensitive to those who may have difficulty sharing in social settings. Others may be unable to identify with a given situation or issue. Additionally, effectively broaching difficult topics tends to involve approaching them from personal experience. The inability to identify with an issue may result in silence (Brookfield & Preskill, 1999 p. 75). For others, silence can provide time for contemplation leading to a more in-depth discussion (Brookfield & Preskill, 1999 p. 226).

Disclosing to others or asking for information from them (receiving feedback directly or indirectly) can be threatening. It is important to be sensitive to the needs of others when we interact with them. We must be accepting ourselves and others. We must also guard against allowing perceptions to bias the interpretation of our observations. As integration and acceptance increases, the desire to give judgmental or evaluative feedback tends to decrease (i.e., pejorative or critical feedback) (Luft, 1969).

The goal of giving and receiving feedback is to move information from the hidden area and UA into the OA where it is available to everyone. This process can move information from the unconscious into the OA as well. Increasing self-awareness may be insightful and inspirational (Hanson, 1973; Luft, 1969). It helps build and reinforce effective working relationships which is the backbone of social inclusion.

SEPARATING THE SOCIAL FROM THE ENVIRONMENT

Activity occurs within a mutual exchange with others and the environment. It is distinct, yet elicits, from others. The other-context exchange is a mutual connectedness (Law, Cooper, Stewart, Rigby, & Letts, 1996, p. 10) which gives rise to the subsummation of one within the other (Gilligan, 2015, p. 75). Bandura speaks to this when he describes people as being subject to "sociostructural influences" whereby they are both "producers and products of social systems — which are devised to organize, guide, and regulate human affairs in given domains by authorized rules and sanctions...created by human activity" (Bandura, 1997, p. 6). We find value in better understanding the difference between the social and environmental. By separating the social interaction from the contexts in which they occur (R. E. Morehouse, Personal Communication, May 17, 2017).

> We can describe the social world as society and its organizations. This includes the notion of individual rank and status. It also includes companionship and human interaction. The environment, on the other hand, can be defined as the setting of conditions in which a particular activity takes place including the overall and unique structures in which persons interact. It is of practical value for our discussion of the AS to separate these constructs, which are often offered as part and parcel with each other (see Fig. 1A for the Accommodation System Model). Both the social and the environment fit into the overall scheme of human development occurring at the same time. They are both separate and united. Separation makes it possible to study their actual relationship as a process" (Valsiner, 1998, p. 21). The concept of agency exemplifies this relationship. For Bandura (1997), agency is an important component of psychological development that evolves gradually in a social environment. This stems from the recognition that one is the agent of one's own actions (Morehouse, in press).
>
> Human development is "possible by the renegotiability of the semiotic constraint system that persons use to organize their personal experiences" (Valsiner, 1998, p. 33). In other words, we humans develop because we can, in part, using a variety of sign systems, organize our world by renegotiating our place in the social world. Further, Harré (Harré & van Langenhove, 1999, p. 16) argues that language is one of the ways persons are positioned hierarchically in social and cultural situations (Morehouse, 2015). It is possible to negotiate and renegotiate the meaning of our experience, if we can create a jointly shared domain within which mutual exchange is possible (Morehouse, 2012). "In contemporary discussions about communication, this domain is labeled intersubjectivity" (Valsiner, 1998, p. 33). For Valsiner, this intersubjectivity is called the social world and is distinct from the environment.
>
> We live in a variety of physical environments including the natural and the humanly constructed worlds. Interwoven throughout is the social world. Both

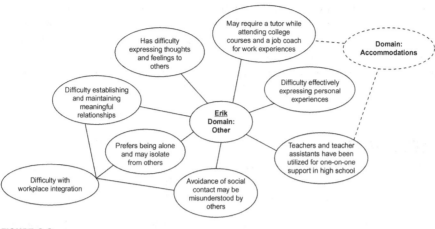

FIGURE 2.3

Erik—case study map: other.

the interpersonal and the environmental influence human development but *in different ways and in different degrees.* The environment, though a part of the whole system of interaction, generally has less influence on the self than does the social world. And traditionally, environments tend to be weighed in one direction—from environment to person (R.E. Morehouse, Personal Communication, May 17, 2017).

The value of considering ourselves and others separately from (but in relation to) our environment, then, allows for a ready discussion of the intricacies associated with providing accommodation and the impact they have on the tasks being performed, in specific contexts, and on the interpersonal negotiations that necessarily occur. All of which impact our being ready, willing, and able to integrate successfully realize a vocational identity. Fig. 2.3 illustrates a case study map for Erik regarding consideration of the social domain.

SUMMARY

The other represents the social dimension of being in the workplace and is one of the five domains comprising the AS which also includes Self, Environment, Accommodations, and Situated Experience. Each of the domains reciprocally interact with the others. We can view others as selves in their own right. As a result, communication, language, and dialog are foundational to the development of workplace relationships between the self and others. True integration and inclusion within the workplace is a process borne of mutuality and an awareness of self and others. We mutually construct meaning in the workplace through a negotiation of shared experiences. Unlike similar models (Bailey, 1982; Bandura, 1978; Bronfenbrenner, 1986; Cook & Polgar, 2012), the AS separates the social and environment constructs as a means to better delineate the service delivery process. Separating these constructs also emphasizes the fundamental importance the social domain plays in successful workplace integration and social inclusion.

REFERENCES

Bailey, R. W. (1982). *Human performance engineering: A guide for system designers.* Upper Saddle River, NJ: Prentice Hall Professional Technical Reference.

Bandura, A. (1978). The self system in reciprocal determinism. *American Psychologist, 33* (4), 344−358.

Bandura, A. (1997). *Self-efficacy: The exercise of control.* New York, NY: W. H. Freeman and Company.

Bronfenbrenner, U. (1986). Ecology of the family as a context for human development: Research perspectives. *Developmental Psychology, 22*(6), 723.

Brookfield, S. D., & Preskill, S. (1999). *Discussion as a way of teaching* (Vol. 85). San Francisco, CA: Jossey-Bass.

Cook, A. M., & Polgar, J. M. (2012). *Essentials of assistive technologies.* St. Louis, MO: Mosby.

Deci, E. L. (1995). *Why we do what we do: Understanding self-motivation.* New York, NY: Penguin Books.

Drucker, P. F. (2010). *Managing oneself. Harvard business review, on managing oneself* (pp. 13−32). Boston, MA: Harvard Business Review Press.

Erikson, E. (1998). *The life cycle completed.* New York, NY: W. W. Norton Company, Inc.

Freire, P. (2000). *Pedagogy of the oppressed.* New York, NY: Bloomsbury Publishing.

Gilligan, C. (2015). The listening guide method of psychological inquiry. *Qualitative Psychology, 2*(1), 69−77.

Hanson, P. C. (1973). *The Johari window: A model for soliciting and giving feedback. The 1973 annual handbook for group facilitators.* San Diego, CA: Pfeiffer & Company.

Harré, R., & van Langenhove, L. (1999). The dynamics of social episodes. In R. Harré, & L. van Langenhove (Eds.), *Positioning theory: Moral contexts of intentional action* (pp. 1−13). Hoboken, NJ: Wiley-Blackwell.

Josselson, R. (1992). *The space between us: Exploring the dimensions of human relatedness.* San Francisco, CA: Jossey-Bass Inc. Publishers.

Josselson, R. (2003). The space between in group psychotherapy: Application of a multidimensional model of relationship. *Group, 27*(4), 203−219.

Josselson, R. (2007). *Playing pygmalion: How people create one another.* Lanham, MD: Rowman & Littlefield Publishers, Inc.

Juniper, D. (1995). Nine insights for a counsellor-manager. *Employee Councelling Today, 7*(4), 4−14.

Lave, J., & Wenger, E. (1991). *Situated learning: Legitimate peripheral participation.* Cambridge, UK: Cambridge University Press.

Law, M., Cooper, B., Stewart, D., Rigby, P., & Letts, L. (1996). The person-environment-occupation model: A transactive approach to occupational performance. *Canadian Journal of Occupational Therapy*, *63*(1), 9−23.

Lewin, K. (1997). Conduct, knowledge, and acceptance of new values. In G. W. Lewin (Ed.), *Resolving social conflicts & field theory in social science* (pp. 49−55). Washington, DC: American Psychological Association.

Luft, J. (1969). *Of human interaction: The Johari model.* Palo Alto, CA: Mayfield Publishing Company.

Morehouse, R. E. (2012). *Beginning interpretive inquiry: A step-by-step approach to research and evaluation.* New York, NY: Routledge.

Morehouse, R. E. (2015). A case for psychology as a human science. *The Journal of Psychology and Clinical Psychiatry*, *6*(2), MedCrave, 1−4.

Morehouse, R.E. (in press). *Caring thinking, education of emotions, and the community of inquiry: A psychological perspective in Ann Margaret Sharp: A life teaching community.* New York, NY: Routledge.

Schein, E. H. (2013). *Humble inquiry: The gentle art of asking instead of telling.* San Francisco, CA: Berrett-Koehler Publishers, Inc.

Selman, R. L. (2003). *Promotion of social awareness: Powerful lessons for the partnership of developmental theory and classroom practice.* New York, NY: Russell Sage Foundation.

Valsiner, J. (1998). *The guided mind: A sociogenetic approach to personality.* Cambridge, MA: Harvard University Press.

Vygotsky, L. S. (1978). Interaction between learning and development. In M. Cole, V. John-Steiner, S. Scribner, & E. Souberman (Eds.), *Mind in society: The development of higher psychological processes* (pp. 79−91). Cambridge, MA: Harvard University Press.

Wenger-Trayner, E., & Wenger-Trayner, B. (2015). Learning in a landscape of practice: A framework. In E. Wenger-Trayner, M. Fenton-O'Creevy, S. Hutchinson, C. Kubiak, & B. Wenger-Trayner (Eds.), *Learning in landscapes of practice: Boundaries, identity, and knowledgeability in practice-based learning* (pp. 13−29). New York, NY: Routledge.

Wicklund, R. A., & Gollwitzer, P. M. (1982). *Symbolic self-completion.* New York, NY: Routledge.

The accommodation System: Environment

Anthony Shay

Capacity Building Specialist, Assistive Technologist, and Rehabilitation Specialist, University of Wisconsin-Stout Vocational Rehabilitation Institute (SVRI), Menomonie, WI, United States

The environment is the context within which work occurs. An environment can have a significant impact on people with disabilities. Contexts must be considered as employment goal planning takes place. Environments are dynamic. They are comprised of the artifacts of life including nature, the built environment, and the cultural artifacts of life including the rules, regulations, and politics of engaging with those around us. Job accommodations attempt to bridge the gap between the self and a specific activity caused by a disability within a specific work environment.

ENVIRONMENTAL ALTERIDAD

Giambattista Vico, a 17th century philosopher, posited the relationship between cognition and action as being inseparable components of the self; the connection between knowing and doing were indivisibly interrelated. He further surmised that the interaction of the self with the environment was a driver of the creative process (Hermans, Kempen, & Van Loon, 1992, pp. 23−24). What Vico was referring to was that the self engages in activity within specific contexts, and that the reciprocal interaction between them leads to generativity. Mutual interaction changes both the person and the environment. Each must negotiate the iterative change process.

The Spanish word alteridad, which has no effective English translation, is described as being a "cross between change and difference" (Giminez, 1997, p. 97). All you have to do is view a picture of a work context from the mid-20th century to get a feel for environmental alteridad (see Fig. 3.1); you get a sense of how environments can change and differ from how you experience them now (if not just from the look and feel alone). However, the time and impact of change on contexts is variable. Contexts can change quickly as a matter of business needs, or it can make an environment we have grown accustomed to over a long period of time present very different challenges following an acquired disability. For people with disabilities, the world of work readily reflects environmental

Assistive Technology Service Delivery. DOI: https://doi.org/10.1016/B978-0-12-812979-1.00003-5

FIGURE 3.1

Environmental alteridad.

alteridad. Environments are far from being static entities. The element of time, as in the future orientation of goal setting, duration of tasks and activities, developmental processes, or as defined by the occurrence of events, enhances this dynamism. We change, the environment changes, and these differences must be accounted for if we are to become and remain vocationally competent (Vygotsky, 1994).

CHARACTERIZING ENVIRONMENTS

To effectively engage the environment, we must recognize it as a source of human development rather than merely being the setting in which it occurs (Vygotsky, 1994, p. 349). As dynamic and complex as they are, environments can foster or impede our progress (Cohen & Piper, 2000). How do we characterize an environment? The International Classification of Functioning Disability and Health (ICF) defines environmental factors related to disability. Environmental factors "make up the physical, social, and attitudinal environment in which people live and conduct their lives". They also reflect whether these factors represent a barrier or facilitator and how impactful they are on an individual's functioning (WHO, 2001, p. 171). The models discussed in this chapter consider the social domain as part and parcel with the environment. As discussed previously, the accommodation system views these domains separately. We present them here to maintain the integrity and scope of the environmental models we are presenting in this chapter. The ICF environmental factors include

- Physical (natural) environment and human adaptations: Physical geography, population characteristics, natural plants and animals, climate, natural and man-made events, lighting sources, natural and man-made temporal changes, sound phenomena, vibration (motion), and air quality (pp. 182–186).

- Social (support) relationships: Immediate and extended family, friends, familiar others (peers/colleagues), authority figures, subordinates, personal care providers, strangers, domesticated animals, healthcare professionals, and other professionals (pp. 187–189).
- Attitudes (related to): Immediate and extended family; friends; familiar others; authority figures; subordinates; personal care providers; strangers; societal attitudes; and social norms, practices, and ideologies (pp. 190–191).
- Services, systems, and policies: Services and systems for the production of consumer goods; architecture and construction services; open-space planning, housing; utilities; communication; transportation; civil protection; legal; associations and organizations; media; economic; social security; general social support; health services; education and training services; labor and employment services; and political services.

Context as referred to in the Human Activity Assistive Technology (HAAT) Model consists of four elements:

- Physical: Natural, built surroundings, location of tasks in relation to others, and physical parameters.
- Social: Peers, strangers, socioeconomic status, and learning style.
- Cultural: Social norms derived by a group with means to transfer them.
- Institutional: Formal legal, legislative, sociocultural (Cook & Hussey, 2008, p. 522; Cook and Polgar, 2012, p. 26).

The HAAT Model also suggests three levels of environment. These are the micro, meso, and macroenvironments:

- Microenvironment: The most immediate context such as work, home, and school.
- Mesoenvironment: A more intermediate context such as shopping centers, church, and community facilities.
- Macroenvironment: A more peripheral context includes the political, legal, and behavioral sociocultural environment (Cook and Polgar, 2012, p. 26).

To these levels of environment, psychologist Urie Bronfenbrenner adds two additional elements, drawn from ecological theory: exosystem and chronosystem:

- Exosystems: Environmental settings that have an indirect impact on an individual. The settings are linked in some way with one setting directly involving an individual while other settings represent indirect connections. These typically include a parent or significant other's place of employment, the extended networks of those close to us, and the neighborhoods in which we live and work.
- Chronosystems: The passage of time extends beyond an individual's aging into the environment. "A chronosystem is a property of the surrounding environment not only over the normal life course, but across historical time ... changes over the life course in family structure, socioeconomic status,

employment, place of residence, or the degree of hecticness and ability in everyday life" (Bronfenbrenner, 1994, p. 40).

Using what we learn in our corner of the world tends to be a reflection of the larger world. Our interface with the world of work facilitates development and integration within these contexts: "the reference books one uses, the notes one habitually takes, the computer programs and databases one relies upon, and perhaps the most important of all, the network of friends, colleagues, or mentors on whom one leans for feedback, help, advice, even just for company" (Bruner, 1996, pp. 132–151).

As stated earlier, the *other* represents the interpersonal element of task engagement. There is a tendency to view social engagement as being an aspect of the environment (Bailey, 1982; Bandura, 1978; Bronfenbrenner, 1986; Cook & Polgar, 2012). The intimate mutuality between contexts and the social activity which occurs in them gives rise to this tendency. Others represent individuals to the same extent as the self. Each individual has a need to identify with the responsibilities and contributions of every other individual on a worksite. Indeed, we have a basic human desire for interaction with others which provides a strong motivation to find and maintain employment. Collective effort and mutuality are drivers of vocational task engagement. Separation of the environmental and social domains improves their understanding and how interpersonal enablement has an impact on job maintenance.

CONTEXT-DEPENDENT LEARNING

Situated learning is context dependent. It is a complex interplay of the self, the environment, and the tasks performed—with a focus on tasks in the plural: "a shift away from a theory of situated activity in which learning is reified as one kind of activity, and toward a theory of social practice in which learning is viewed as an aspect of all activity" (Lave & Wenger, 1991, pp. 37–38). Learning is context dependent through "elaborating existing frames of reference, learning new frames of reference, transforming frames of reference or points of view, and transforming habits of mind" (Kasl & Elias, 2000, p. 231). We develop through effective engagement with an increasingly complex environment (Kasl &Elias, 2000). A situated experience is not necessarily bounded by space and time but by the culture of the social group and through the development of identity (Kasl & Elias, 2000; Lave & Wenger, 1991), the building of a sense of fellowship and belonging. Situated learning experiences may begin as apprenticeships, work studies, work/volunteer experiences or similar. It is through the process of gradual exposure to work contexts that we develop expertise and a context-dependent identity. We learn from and with others in both formal and informal activities from those with whom we have direct and indirect contact, and as both passive and active members of our work community (Wenger, White, & Smith, 2009, pp. 6–8).

We learn in relation to others within the environments that work occurs, using the tools of the trade, with accommodations in place. The subsequent quality of our situated experiences reinforces motivation and task-oriented behavior. Our ever-maturing sense of identity within the work environment begins vicariously in most cases. It builds toward our sense of belonging to and membership within a vocational community (Gollwitzer & Kirchhof, 1998; Lave & Wenger, 1991).

CHALLENGING ENVIRONMENTS

Environments "occur outside individuals and elicit responses from them." Effective responses require us to consider "contextual influences, temporal factors, and [our] physical and psychological characteristics." Our contexts are continually shifting and as contexts change, the behavior necessary to accomplish a goal also changes" (Law, Cooper, Stewart, Rigby, & Letts, 1996, p. 10). Overcoming environmental challenges requires a process by which we dynamize, master, and humanize reality, thereby "giving temporal meaning to graphic space, by creating culture." We have the ability through informed choice to direct action toward manifesting change and control over our environment. Later in this chapter, we outline the contextual functional skills domain typically used in assistive technology assessment. Disability drives a renegotiation (i.e., rehabilitation) between an individual and the contexts in which they operate. Our capacity for self-direction may become impaired. Personality, cognition, and the manner in which the environment is engaged may change. However, we can optimize our environments to provide the greatest opportunity to achieve satisfaction and meaning (Csikszentmihalyi, 2003).

ENVIRONMENTAL OPTIMIZATION

Human nature is a product of the environments we create for ourselves (Schwartz, 2016). From this perspective, human nature is designed more so by others than revealed to us individually. Disability and lower resourced environments can lead to significant functional impairments with a lifelong impact (Bruner, 1966; Feuerstein, Falik, & Feuerstein, 2015). When we are able to effectively self-direct our behavior toward optimizing vocational experiences, we seek higher challenge or skills, the reduction of challenge or skills, or we switch activities entirely to maintain a higher than average balance between challenges and skills in specific contexts and activities. This leads to and reinforces well-being. Environmental optimization, individually or for groups of people, involves modifying or designing environments to maximize conduciveness for the people within them toward an optimal experience—a balance of vocational challenges with extant skills—avoiding over or under stimulation, negative emotions, and poor motivation. Universal design is a type of environmental optimization. It is proactive in that it "anticipates the needs of a diverse group of users" seeking to enable access for a population who requires it for a particular purpose (Burgstahler, 2008, p. 7). An optimized work environment leads to positive situated experiences.

According to Feuerstein and Lewin-Benham, developmental deficiencies due to the environment can be overcome, at times quite significantly, through mediation. This intensive practice of facilitating meaning making within the learning context has the effect of also facilitating the development of intrinsic motivation in individuals (Feuerstein and Lewin-Benham, 2012). This affirms the value of the individual regardless of the etiology of the lack of development an individual brings to rehabilitation efforts. It reinforces a person-centered approach to bridging environmental gaps through the provision of accommodations.

Achieving satisfaction with and meaning in our work life is largely dependent on the ability to discriminate between and choose to attend to stimuli in the environment: "one has to begin selecting stimuli from the surroundings, restrict one's attention to a manageable pattern of items *about which one can do something*" (Csikszentmihalyi, 1975, pp. 191–192). Effort and persistence is the key to maintaining attention. Bruner suggests that individuals engage in three behaviors toward "economizing effort" including focusing attention on "those things that are somehow essential to the enterprises in which we are engaged ... recod[ing] into simpler form the diversity of events that we encounter" and dealing with excessive environmental stimuli through the use of technology such as paper, pencil, and cameras. Bruner states that all of these strategies help reduce cognitive effort although no strategy "can succeed fully" since we are not capable of "knowing everything about a particular situation" (Bruner, 1958, pp. 86–87). Herbert Simons (cited in Perkins, 1981) ascribes to the notion that people do not necessarily strive to attain the highest possible goal but rather to *satisfice*. They "strive until they achieve so much and then stop. They work to satisfy standards of adequacy ... [which] exceeds some threshold of acceptability" (Perkins, 1981, p. 156). As such, achievement may well be relative to the goals we set in relation to our perceived needs. With the prudent investment of our cognitive resources, the consideration of our unique needs, the alignment of necessary accommodations, environmental optimization, and with effective task engagement we stand ready to deliver on the intention of goal achievement (Nakamura & Csikszentmihalyi, 2005). Fig. 3.1 illustrates a case study map for Erik regarding environmental considerations.

CONTEXTUAL SKILLS DOMAIN

- Physical: The physical nature of the environment in which assistive technology will be used (including transit between environs).

- Social: Context dependent availability and type of interaction—human experiences.
  - Opportunity barriers: Barriers that an individual has no control over which impact their capacity for full inclusion and integration in the community. These barriers present in policy, attitudes, KSAs (knowledge/skills/abilities), and standard practices.
  - Access barriers: KSAs, attitudes, and limited resource—including preferences of assistants or aides providing services (or other rehabilitation team/family members).
- Cultural: The capacity for how an individual's upbringing and related influences impact the use of assistive technology—familial and social influences and preferences based on contextual upbringing.
  - Prescriptive sociocultural expectations: The social norms derived by a group with means to transfer them. Cultures prescribe these norms and expectations as a means to streamline shared contextual experiences.
- Institutional: The capacity for context dependent legislative, political, governmental, and regulatory laws, policies and procedures to impact access and use of assistive technology.
  - Practice: Activities typically engaged in by individuals which are institutionally limited by informal practice but not formal restriction.
  - Legislative: Formal, legal barriers to access based on regulatory processes—includes physical and service access.

The Assistive Technology Functional Skill Domains Sections of this text is adapted from Cook, A. M. & Polgar, J.M. (2015). *Assistive technologies: Principles and practice* (4th Edition). St. Louis, MO: Mosby. Used with Permission.

SUMMARY

The environment represents the contexts within which work occurs. It is concerned with the variables that have an impact upon people with disabilities who are in the process of developing a job or seeking to maintain employment. Environments are dynamic entities. Both the self and the environment changes. Others within the environment also change. The environment can enable or hinder function to varying degrees. This can have an impact on the effectiveness or appropriateness of accommodations. Job accommodations attempt to bridge the gap between the self and a specific activity caused by a disability within a specific work environment. When we optimize our work environment we not only improve access to and inclusion within that context, we also stand a greater chance of improving the value we place on the vocational experience. Fig. 3.2 illustrates a case study map for the environmental domain.

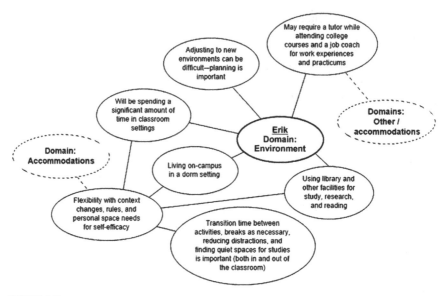

FIGURE 3.2

Erik case study map: environment.

REFERENCES

Bailey, R. W. (1982). *Human performance engineering: A guide for system designers.* Upper Saddle River, NJ: Prentice Hall Professional Technical Reference.

Bandura, A. (1978). The self system in reciprocal determinism. *American Psychologist, 33* (4), 344−358.

Bronfenbrenner, U. (1986). Ecology of the family as a context for human development: Research perspectives. *Developmental Psychology,* 22(6), 723.

Bronfenbrenner, U. (1994). Ecological models of human development. *Readings on the Development of Children, 2,* 37−43.

Bruner, J. (1958). Social psychology and perception. In E. E. Maccoby, T. M. Newcomb, & E. L. Hartley (Eds.), *Readings in social psychology* (pp. 85−94). New York, NY: Henry Holt and Company Inc.

Bruner, J. S. (1966). *Toward a theory of instruction.* Cambridge, MA: Harvard University Press.

Bruner, J. S. (1996). *The culture of education.* Cambridge, MA: Harvard University Press.

Burgstahler, S. E. (2008). Universal design in higher education. In S. E. Burgstahler, & R. C. Cory (Eds.), *Universal design in higher education: From principles to practice* (pp. 3−20). Cambridge, MA: Harvard Education Press.

Cohen, B. C., & Piper, D. (2000). Transformation in a residential adult learning community. In J. Mezirow (Ed.), *Learning as transformation* (pp. 205−228). San Francisco, CA: Jossey-Bass.

Cook, A. M., & Hussey, S. M. (2008). *Assistive technologies: Principles and practice* (3rd Ed). St. Louis, MO: Mosby.

Cook, A. M., & Polgar, J. M. (2015). *Assistive technologies: Principles and practice* (4th Edition). St. Louis, MO: Mosby.

Csikszentmihalyi, M. (1975). *Beyond boredom and anxiety: Experiencing flow in work and play*. San Francisco, CA: Jossey-Bass Inc.

Csikszentmihalyi, M. (2003). *Good business: Leadership, flow, and the making of meaning*. New York: Penguin Books.

Feuerstein, R., Falik, L. H., & Feuerstein, R. S. (2015). *Changing minds & brains: The legacy of Reuven Feuerstein*. New York, NY: Teacher's College Press.

Feuerstein, R., & Lewin-Benham, A. (2012). *What learning looks like: Mediated learning in theory and practice, K-6*. New York, NY: Teacher's College Press.

Giminez, F. (1997). Community, alteridad and difference, the voices of difference. *Analytical Teaching, 17*(2), 77−85.

Gollwitzer, P. M., & Kirchhof, O. (1998). The willful pursuit of identity. In J. Heckhausen, & C. Dweck (Eds.), *Motivation and self-regulation across the lifespan* (pp. 389−423). New York, NY: Cambridge University Press.

Hermans, H. J., Kempen, H. J., & Van Loon, R. J. (1992). The dialogical self: Beyond individualism and rationalism. *American Psychologist, 47*(1), 23−33.

Kasl, E., & Elias, D. (2000). Creating new habits of mind in small groups. In J. Mezirow (Ed.), *Learning as transformation* (pp. 229−252). San Francisco, CA: Jossey-Bass.

Lave, J., & Wenger, E. (1991). *Situated learning: Legitimate peripheral participation*. Cambridge, UK: Cambridge University Press.

Law, M., Cooper, B., Stewart, D., Rigby, P., & Letts, L. (1996). The Person-Environment-Occupation Model: A transactive approach to occupational performance. *Canadian Journal of Occupational Therapy, 63*(1), 9−23.

Nakamura, J., & Csikszentmihalyi, M. (2005). The concept of flow. In C. R. Snyder, & S. J. Lopez (Eds.), *Handbook of positive psychology* (pp. 89−105). New York, NY: Oxford University Press.

Perkins, D. N. (1981). *The mind's best work*. Cambridge, MA: Harvard University Press.

Schwartz, B. (2016). The way we think about work is broken. [Video podcast]. In: Films Media Group, film distributor, & TED conferences LLC. Available at: https://www.youtube.com/watch?v = 3B_1itqCKHo.

Vygotsky, L. S. (1994). The problem of the environment. In R. Van Der Veer, & J. Valsiner (Eds.), *The Vygotsky reader* (pp. 338−354). Cambridge, MA: Blackwell Publishers, Inc.

Wenger, E., White, N., & Smith, J. D. (2009). *Digital habitats: Stewarding technology for communities*. Portland, OR: CPsquare.

World Health Organization. (2001). *International Classification of Functioning Disability and Health*. Geneva: Author.

Accommodation System: Accommodations

4

Anthony Shay

Capacity Building Specialist, Assistive Technologist, and Rehabilitation Specialist, University of Wisconsin-Stout Vocational Rehabilitation Institute (SVRI), Menomonie, WI, United States

THE ACCOMMODATION SYSTEM: ACCOMMODATIONS

When we have a disability, accommodations allow us to engage in activities we might otherwise have great difficulty with or be unable to perform. They are directly or indirectly extrinsic enablers of activity. They come in all shapes and sizes and are used in as many environments. The accommodations domain is one of five which comprises the accommodation system (AS): self, other, environment, accommodations, and situated experience. Each of the domains reciprocally interacts with the others. The overarching goal of providing accommodations is to empower individuals with disabilities, to build self-reliance, and to promote full inclusion in society. Accommodations facilitate task engagement, increase well-being, are drivers of the development and maintenance of meaningful relationships, and help us negotiate our surroundings.

WHAT IS AN ACCOMMODATION?

Barriers to integrated employment for people with disabilities and the need for accommodations to facilitate integration are readily recognizable in the workforce (Anderson et al., 2015). The term *accommodation* refers to any strategy we may use to facilitate overcoming a functional limitation toward engaging in an activity within a specific environment. According to the United States Equal Employment Opportunity Commission (2018) an accommodation is "any change in the work environment or in the way things are customarily done that enables an individual with a disability to enjoy equal employment opportunities." Accommodations have also been called compensatory strategies (Wehman, Inge, Revell, & Brooke, 2007). Considerations of the purpose for, and access to, accommodations are keys to their effectiveness (Elliott, Kurylo, & Rivera, 2005). According to the Job Accommodation Network (2018, p. 3), the provision of accommodations should be done on a "case-by-case basis" due to our unique needs as individuals including "limitations, restrictions, specific accommodation needs, and the impact accommodation will have on job performance and business operations."

Assistive Technology Service Delivery. DOI: https://doi.org/10.1016/B978-0-12-812979-1.00004-7

An accommodation is an intervention. It is a strategy developed not to "fix" disability but rather to enhance ability. Cook and Hussey (2008, p. 93) state that the purpose of an intervention is not to rehabilitate a consumer but to enable them in functional activities. Meteer and Sira find "the nature and degree of impairments, the resulting disabilities, and the environmental context" are the determining factors in choosing an effective accommodation. Matching an accommodation to our individual needs is paramount. Meteer and Sira (2006, p. 325) further advise, "Be creative. Obtain knowledge and experience in the area, and be prepared to adopt an eclectic, problem solving approach." Intervention strategies work best when we perceive them to be a valid and reliable means of overcoming the challenges we face in the workplace. Pagan Kennedy in her book *Inventology* goes as far as to say that innovation is not a cognitive process of the innovator. Rather, it is a process revealed through real-world usage of the innovation (Kennedy, 2016, p. 101). A truly effective accommodation is a product of our need for overcoming our limitations in the real-world. We must factor in aspects of the self, others, environment, and how these all interact to meet the need. A positive situated experience can be summed up by what Kennedy refers to as the "Pong Effect":

- The innovator creates based on a perceived need (sometimes a rather clunky something or other that is less than perfect but is still useful);
- People use the creation;
- They recognize the value in it;
- They build it into their lives;
- They develop habits around its use; and then
- They find they cannot live without the creation.

Transformational innovation like this have what Kennedy dubbed a "pong-i-ness"—a stick-to-itiveness that keeps people psychologically invested in a new device or system until all the kinks have been worked out (Kennedy, 2016, pp. 105–106).

THE ROLE OF PREINTERVENTIONS

The integrated multiintervention paradigm for assessment and application of concurrent treatments (IMPACT2) model provides a theoretical framework for conceptualizing the relationship between context, assistive technology (AT) interventions, and outcome variables. This model has six basic components when preinterventions are considered:

- Preinterventions: Include health promotion and universal design (UD). This component remains a consideration throughout the AT intervention process.
- Context: Includes person, task, and environment.
- Baseline: A measurement of current functional capacity of the individual prior to an intervention approach.

- Intervention approaches: Six different modes of intervention which include
  - Reducing the impairment
  - Compensating for the impairment
  - Using AT
  - Task or contextual redesign
  - Using a personal assistant
- Proxy outcomes: Includes precursor variables which are a correlated subset of the proxy outcomes which are a correlated subset of the intervention outcomes proper. Proxy outcomes are often confused with intervention outcomes proper
- Outcomes: Measurement of functional capacity and compared against baseline data to determine the effectiveness of the intervention.

The IMPACT2 model considers each of the components in turn while keeping preinterventions in mind with the outcome ideally being an increase in functional capacity. It assumes a person-centered approach to addressing an individual's needs: "Clearly, assistive technology devices and services pose a primary potential intervention. Following these interventions, we assume that the initial baseline function would be improved. This would be measured as outcomes subsequent to the intervention" (Schwanke & Smith, 2005, p. 197).

The IMPACT2 model identifies as preinterventions health promotion and UD under the heading of "context" (Elsaesser & Bauer, 2011). Preinterventions affect human activity output in context with the effect of minimizing or eliminating the need for an AT intervention: "By definition, a successful preintervention would result in an elevated baseline function, so a lower intensity or omission of a later intervention would be possible" (Schwanke & Smith, 2005, p. 197). These occur prior to functional declines and the need for AT. Preinterventions are considered throughout the process. Preinterventions may provide benefits in terms of supporting educational activities facilitating awareness and reinforcing motivation (Öberg, 2008) in lieu of, in advance of, or in service of AT activities. The IMPACT2 model provides a useful way to conceptualize our satisfaction with the service we receive and the outcomes providers track. The use of the IMPACT model in the discipline of AT service provision instantiates its adaptation to the field of employment for people with disabilities and accommodations more generally.

Our capacity to overcome functional limitations in the workplace can be viewed as both an aspect of health behavior and learning to overcome limitations through the intervention process. According to Glanz, Rimer, and Lewis (2002, p. 10), "In the broadest sense, *health behavior* (emphasis in the original) refers to the actions of individuals, groups, and organizations as well as their determinants, correlates, and consequences, including social change, policy development and implementation, improved coping skills, and enhanced quality of life." This behavior is both subjectively and objectively quantifiable. Subjective behavior includes our innate qualities (e.g., beliefs, goals, motivation, strengths, perceptions, feelings, and personality traits) and objective behavior includes our observable behavior (e.g., habits, actions, nonverbal behavior, health maintenance, and improvement activity) (Glanz et al., 2002, p. 10).

Closely related to the above personal attributes competence, autonomy, and relatedness "when satisfied yield enhanced self-motivation and mental health and when thwarted lead to diminished motivation and well-being" (Ryan & Deci, 2000, p. 68). Selection of an appropriate intervention, one that supports and fosters both the subjective and objective aspects of the self in the workplace, is a necessary component of effective service provision. A proper match between the individual and the job goal (including job tasks, work contexts, and relationship building and maintenance) may result in minimal or no required interventions. In this sense, employment (i.e., the *type* of job selected) may represent both an intervention and an outcome. We offer many examples of accommodations, interventions, and compensatory strategies following this chapter. The following discussion groups compensatory strategies into three broad categories: individualized accommodations, AT, and UD.

Individualized accommodations. The Americans with Disabilities Act (ADA) defines the provision of vocational accommodations for individuals with disabilities: In general, an accommodation is any change in the work environment or in the way things are customarily done that enables an individual with a disability to enjoy equal employment opportunities. There are three categories of "reasonable accommodations":

1. Modifications or adjustments to a job application process that enable a qualified applicant with a disability to be considered for the position such qualified applicant desires; or
2. Modifications or adjustments to the work environment, or to the manner or circumstances under which the position held or desired is customarily performed, that enable a qualified individual with a disability to perform the essential functions of that position; or
3. Modifications or adjustments that enable a covered entity's employee with a disability to enjoy equal benefits and privileges of employment as are enjoyed by its other similarly situated employees without disabilities (ADA.gov, 1990).

Equal employment opportunities are those which allow an individual with a disability to attain the same level of performance or to enjoy equal benefits and privileges of employment as are available to an average similarly situated employee without a disability (Job Accommodation Network, 2017).

We can describe accommodations by the qualities they possess: the degree of technology they employ (e.g., no, low, mid, or high technology); how expensive they may be; how difficult they are to find, produce, or create; how portable they may be; the skill required to learn to use them, including our tolerance for their use (e.g., gadget tolerance); the ability to adapt them for use in an intervention system (e.g., computer, accessories and software, use with a job coach, adding curb cuts at work); whether the accommodation is replacing, augmenting, or supplementing functions; whether the accommodation can be used for more than a single purpose; and whether a service or item is available commercially or if it needs to be carved-out from existing services or developed especially for an individual's need (e.g., training coworkers regarding disabilities in the workplace) (Cook & Polgar, 2012; Kurtts, Dobbins, & Takemae, 2012). Virtually, any strategy can potentially be an accommodation when designed to facilitate task engagement. People may also serve in this capacity (e.g., job coaches, interpreters, personal assistants). Most workplace accommodations tend to be "simple and involve minimal cost" reflecting four qualities: they are effective, transparent (i.e., they either have a positive or no impact on a business' operations), timely, and durable (i.e., they offer flexible use throughout a person's employment tenure) (Meuller, 2011). Once accommodations have been selected addressing our unique needs, they are implemented within the work context.

When seeking a reasonable accommodation, we must inform the employer, verbally or in writing, at the time we realize the need for an accommodation due to disability. This means we cannot withhold the information until an employer questions why work tasks are not being completed as expected. We may be required to provide medical documentation to verify the need for an accommodation. The only factor mitigating the provision of an accommodation by an employer is whether it would result in an undue hardship. Undue hardships may be the result of an accommodation request creating significant difficulty or expense for an employer (Job Accommodation Network, 2017).

Assistive Technology (AT). Langton and Ramseur find that accommodations and AT are indispensable to meeting the expectations of people with disabilities. Although they "do not hold all of the answers to the employment needs of individuals [with disabilities] ... achieving successful employment outcomes ... often cannot be achieved without the accommodations and performance enhancements that AT can provide" (Langton & Ramseur, 2001, p. 36). Technology is a critical factor in workplace integration and task mastery. Its importance cannot be overstated (Lave & Wenger, 1991). The use of AT is universally applicable across peoples regardless of the type of disability, severity of functional limitations, or the contexts in which AT is used (Smith & Scherer, 1998). Later in this chapter, we present the assistive technology functional skills domain typically used in an AT assessment. The United States Assistive Technology Act of 2004 defines AT, an AT device, and AT services as

- AT: The term "assistive technology" means technology designed to be utilized in an AT device or AT service.
- AT device: The term "assistive technology device" (sometimes referred to as *aids and devices*) means any item, piece of equipment, or product system, whether acquired commercially, modified, or customized, that is used to increase, maintain, or improve functional capabilities of individuals with disabilities.
- Assistive technology service: The term "assistive technology service" means any service that directly assists an individual with a disability in the selection, acquisition, or use of an AT device.

AT service delivery includes AT needs assessments and functional evaluations, acquisition of AT, matching AT to individual needs, maintenance and repair of AT, inclusion of associated services (e.g., education, therapies, programs), training and technical assistance (e.g., for family, guardians, advocates, representatives, professionals, employers, AT vendors, and rehabilitation providers), and for expansion and access of AT services to people with disabilities (Pub. L. No. 108-364, 2004).

Universal Design (UD) and individual accommodations (IA). UD and IA both serve to facilitate overcoming a functional limitation toward engaging in an activity within a specific environment. The difference is found in whether the intervention is person-centered and individualized or whether it is intended to address the more generalized needs of as large a group of people as possible. The former is an individualized accommodation. The latter is more process-oriented accessible technology known as UD (Follette Story & Mueller, 2011). The focus of UD efforts is to provide equal access to everyone without regard to individualized accommodations (Adya, Samant, Scherer, Killeen, & Morris, 2012).

Universal Design. UD is also known as "barrier-free design, accessible design, inclusive design, and design-for-all" (Ostroff, 2001, p. 1.5) and involves the design, development, and production of environments, buildings, and products for use by the population in general. This concept is contrasted by the acquisition of commercially available products for use as-is, or as modified, or through the fabrication of custom products for use by an individual or small number of individuals (see the discussion above regarding accommodations and AT) (Burgstahler, 2017). There are differences between UD and IA.

UD and IA both serve the purpose of facilitating an individual's interface with their environment. The difference between these two concepts is that IA focus on a single person's interface with their environment and, as stated earlier, UD seeks to provide ease of access to the environment for people in general (Ostroff, 2001). The difference between AT and UD is a function of the individual and the activity which is the focus of the need for the accommodation. The situated experience dictates the perception of the accommodation and thus the designation of AT or UD (Follette Story & Mueller, 2011). The focus of UD efforts is to provide equal access to everyone without regard to disability (Adya et al., 2012).

Follette Story and Mueller (2011, pp. 32.2−32.5) offer a UD model composed of seven principles:

- Equitable use:
  - The design is useful and marketable to people with diverse abilities.
- Flexibility in use:
  - The design accommodates a wide range of individual preferences and abilities.
- Simple and intuitive use:
  - Use of the design is easy to understand, regardless of the user's experience, knowledge, language skills, or current concentration level.

- Perceptible information:
  - The design communicates necessary information effectively to the user, regardless of ambient conditions or the user's sensory abilities.
- Tolerance for error:
  - The design minimizes hazards and the adverse consequences of accidental or unintended actions.
- Low physical effort:
  - The design can be used efficiently and comfortably and with a minimum of fatigue.
- Size and space for approach and use:
  - Appropriate size and space are provided for approach, reach, manipulation, and use regardless of the user's body size, posture, or mobility.

Smith proposed a new UD model composed of eight principles. According to Smith, these principles reflect a more fundamental nature of UD as more incremental and an ongoing work-in-progress rather than UD always being just out of reach for users. The eight principles of Smith's UD model includes

- Body fit—accommodating a wide a range of body sizes and abilities
- Comfort—keeping demands within desirable limits of body function and perception
- Awareness—insuring that critical information for use is easily perceived
- Understanding—making methods of operation and use intuitive, clear, and unambiguous
- Wellness—contributing to health promotion, avoidance of disease, and prevention of injury
- Social integration—treating all groups with dignity and respect
- Personalization—incorporating opportunities for choice and the expression of individual preferences
- Appropriateness—respecting and reinforcing cultural values together with the social and environmental context of any design project.

With these principles in mind, Smith (2011) proposes a new definition of UD: "Universal design is a process that increases usability, safety, health and social participation, through design and services that respond to the diversity of people and abilities."

Usability and accessibility. Both of these terms refer to the fulfillment of an individual's functional requirements with accessibility referring also to the measurement of the fulfillment against an established standard (e.g., ADA, the Individuals with Disabilities Education Act, United Nations Convention on the Rights of Persons with Disabilities). The term accessibility tends to focus largely at the level of the individual, whereas usability tends toward a more global focus in addressing how functional an approach is in meeting the needs of people with disabilities. The cost, access to funding, and affordability all have an impact on access to accommodations for those who use them (Brewer, 2011).

Usability is comprised of five attributes: learnability, efficiency, memorability, errors, and satisfaction. When we refer to usability, we address user-centered principles such as the need for user interfaces to be easy to learn, amenability to practical application, intuitive recall, minimization of the consequences of error, and meaningful usage. Of note are the nuances in the definitions of some related terms: utility, usability, and useful:

- Utility refers to whether something possesses necessary features (as opposed to it's being inutile or of no use or value to the user);
- Usability refers how satisfying these features are;
- The usefulness of what is being used is a product of both utility and usability—having utility and meeting the expectations and needs of the user (Nielsen, 2017, pp. 3—5).

Fig. 4.1 illustrates a case study map for Erik regarding accommodation needs.

ASSISTIVE TECHNOLOGY SKILLS DOMAIN

- AT characteristics: The overall properties of the technology—larger categorization.
  - Feature: Features include the qualities of the characteristic (the wheel on a wheelchair is a characteristic, a feature would be off-road, wide, green, and camber angle degree of the tires).
  - Feature matching: The process of matching device characteristics/features and an individual's skills and abilities.
- Human-technology interface: This interface includes the conjunctive aspects of the AT with the individual using the technology—that part of the AT with which the individual directly interacts—includes the physical attributes of the AT (dimensions, weight, texture, hardness, display elements) and sensory feedback.
- Quantity of inputs: The number of operational inputs necessary to use the AT device.
- Selection method: Process for determining the best fit for input selection as direct or indirect:
  - Direct selection: Increased physical ability and decreased cognitive ability.
  - Indirect selection: Decreased physical ability and increased cognitive ability.
- Selection set: The total number of selections available for AT device inputs—includes the features of the selection input: size, format, and symbolic representation.

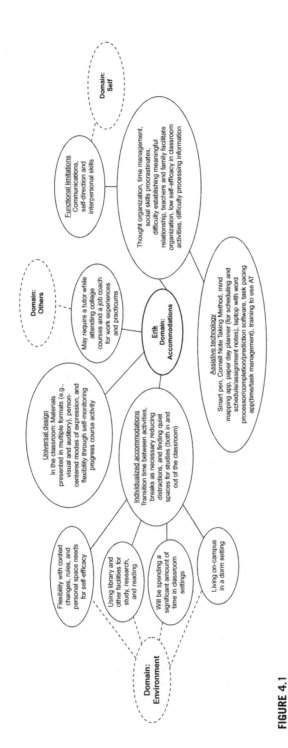

FIGURE 4.1

Erik—case study map: accommodations.

- Processor: The interface component controlling direction of the AT inputs toward desired activity outputs—can be simple mechanical to complex electronics.
  - Directive set size: Overall total number of commands available.
  - Activity output adjustment: Changes made to input based on feedback from output (such as powered wheelchair speed or direction).
- Activity output: The context-based task outcome which precipitated the use of the AT.
  - Magnitude: The degree or effect size of the expected output from the AT (e.g., loudness, brightness, speed, torque, force).
  - Accuracy: The degree to which the activity outcome achieves expected results.
  - Flexibility: The degree to which the AT can be applied in different contexts and to perform different tasks or activities.
- Environmental interface: The sensory interface component of the AT which serves to input context-based sensory data (e.g., pressure sensor, camera, etc.).
  - Range: The upper and lower limits of the sensory input component of the AT.
  - Threshold: The lowest limit of the sensory input component of the AT necessary for effective use of the technology—differs based on need, skill/ability, type of technology, expected activity output, and context.
- Reliable access: The attributes of the AT which lends itself to dependable access—includes mounting, positioning, esthetics, portability, storage, and power (i.e., performance reliability).

The Assistive Technology Functional Skill Domains Sections of this text are adapted from Cook, A. M. & Polgar, J.M. (2015). *Assistive technologies: Principles and practice* (4th Edition). St. Louis, MO: Mosby. Used with Permission.

SUMMARY

The accommodation domain, within the AS model, reflects an active intervention whereby accommodations are matched to an individual's unique needs. Individualized accommodations, AT, and UD interventions, including the training necessary for their effective use, are designed to bridge the gap caused by disability between the performance of specific work activities within specific work contexts. The accommodations domain is one of five which comprises the AS. These domains (self, other, environment, and situated experience) interact dynamically with each other. Preintervention strategies and job choice may mediate the need for an intervention. Accommodations must possess usability and accessibility to effectively accommodate the needs of people with disabilities.

EXAMPLES OF ACCOMMODATIONS, INTERVENTIONS, AND COMPENSATORY STRATEGIES

1. Ability to Opt-Out of Work-Related Social Functions
2. Accommodations Review (Determine, Obtain, Implement, and Review on a Regular Basis)
3. Achievement Testing
4. Additional Time On-Task
5. Address Disability Disclosure and Confidentiality in the Workplace
6. Address Workplace Bullying and Harassment
7. Advocacy/Self-Advocacy Training
8. Affirmations (and Social Validation)
9. Apprenticeship
10. Assign Workplace Disability Rights Champions
11. Assistive/Rehabilitation Technology
12. Assistive/Rehabilitation Technology Training
13. Augmentative and Alternative Communication Devices
14. Behavioral Planning
15. Behavior Management
16. Benefits Analysis
17. Career Day (Career Awareness) Events
18. Changing Supervision Techniques (and Expectations)
19. College/University/Vocational Entrance Examinations (to Determine Interest, Motivation, Individual Needs)
20. Communication Access Real-Time Translation (CART) Services
21. Comprehensive Vocational Assessment
22. Concept Mapping of Work Variables
23. Concurrent Referral to a Disability-Employment Agency and an Employment Only Agency
24. Consultation Services
25. Coworker Disability Training
26. Creating (Reviewing/Maintaining) Safe Work Environments
27. Customized Employment
28. Development of Work Portfolios (e.g., Visual Resume)
29. Developing Task-Specific Routines (Behavioral Strategy)
30. Disability Awareness/Sensitivity Training
31. Disability Skills Training
32. Driver Services for the Blind
33. Employee Awareness Programs (Raising Awareness of them)
34. Employer Visits and Tours
35. Employer/Expert Informational Interviews
36. Entrepreneurial/Inventors Club Memberships
37. Environmental Changes (Lighting, Air Quality, etc.)
38. Ergonomic Assessment
39. Errorless Learning (Unconditional Acceptance)
40. Establishing Bank Accounts
41. Financial Planning/Credit Counseling
42. Flexible Work Hours Scheduling
43. Graduated Return to Work
44. Job/Career Club
45. Job Coaching
46. Job Culture Awareness Training
47. Job Placement

(Continued)

EXAMPLES OF ACCOMMODATIONS, INTERVENTIONS, AND COMPENSATORY STRATEGIES (CONTINUED)

48. Job Shadow(s)
49. Job Sharing (with One or More Additional Workers)
50. Job-Site Coworker Relationship Building
51. Job Task Deadline Flexibility
52. Job Task Modification
53. Job Training Modification
54. Incidental Counseling (Informal and Untrained Others)
55. Independent Living Assessment
56. Individualized Training Programs
57. Individual Placement & Support (IPS)
58. Informal Observations (Watching at a Distance Jobs Being Performed)
59. Input from Interdisciplinary/Rehabilitation Teams (and Stakeholder Groups)
60. Interpreter
61. Labor Market Assessment
62. Low Vision Aides
63. Maintenance (Rent/Room/Board)
64. Mental Health Counseling
65. Mentorship
66. Meta-Cognitive/Self-Regulatory Strategies
67. Mobility & Orientation Training
68. Monitoring Time-Off Activities
69. Neuropsychological Assessment
70. Note/Minute Taker
71. Occupational/Physical Therapy Consultation
72. Off Job-Site Job Coaching and Employment Planning
73. On-the-Job Assessment
74. On-the-Job Training
75. Parent-Teacher Career Development Conferences
76. Passive/Active External Aids for Memory
77. Peer Assistants/Peer Support
78. Personal Assistance
79. Planned Medical Absences from Work
80. Positive Reinforcement
81. Postemployment Services
82. Presage Organization Changes
83. Professional Workplace Visit
84. Progress Measures Creation and Review
85. Provide Communication Alternatives
86. Psychological Assessments
87. Regular Feedback Meeting with Management
88. Remedial Training
89. Resource Ownership (Owning Tools and Equipment Related to Work Activity)
90. Resource & Referral
91. Responsibility Checklist
92. Restorative Practices
93. Role-Playing
94. Scheduled Telephone Calls
95. Secondary School Clubs and Extracurricular Activities

(Continued)

EXAMPLES OF ACCOMMODATIONS, INTERVENTIONS, AND COMPENSATORY STRATEGIES (CONTINUED)

96. Selective/Sustained Attention Activities
97. Self-Management Training
98. Self-talk/Self-belief Therapy (Cognitive Behavioral)
99. Service Learning/Volunteerism
100. Services to Family Members (Child Care)
101. Shifting Supervisory Responsibilities (to a New Supervisor)
102. Shifting Work Duties/Tasks to a Different Location
103. Supervisor/Management Disability Training
104. Supported Employment (and Long-Term Supports)
105. Task Shifting or Exchange with Coworkers
106. Telework/Work from Home
107. Temporary Work/Internship
108. Testing for Adult Basic Education
109. Transition Planning (School-to-Work)
110. Transportation Services
111. Universal/Accessible Design
112. Validation of Limitations Concerns and Feelings
113. Verbal Mediation Strategy
114. Vicarious Learning (Watching the Service Delivery Professional or Others Perform Tasks)
115. Virtual Reality Learning Environments
116. Vocational Assessments
117. Vocational Counseling
118. Vocational Futures Planning
119. Vocational Training
120. Wellness Management Training
121. Work Adjustment Training
122. Work-Related Materials & Tools (Provision of)
123. Work Samples
124. Workstation Assessment
125. Youth Apprenticeship

REFERENCES

Adya, M., Samant, D., Scherer, M. J., Killeen, M., & Morris, M. W. (2012). Assistive/rehabilitation technology, disability, and service delivery models. *Cognitive Processing, 13* (1), 75−78.

ADA.gov (1990). Americans with Disabilities Act of 1990, as Amended. Pub. L. No. 101-336, 104 Stat. 328. Retrieved from https://www.ada.gov/pubs/adastatute08.htm#12111.

Anderson, C., Matthews, P., Hartman, E., Lubinsky, C., Fine, C., Lui, J., & Hartwig, C. (2015). Perceptions on the availability & outcomes of employment services for individuals with disabilities in Wisconsin. Menomonie, WI. Retrieved from: www.uwstout.edu/svri.

Pub. L. No. 108-364 (2004). Assistive Technology Act of 1998, as Amended. Pub. L. No. 108-364. Retrieved from https://www.congress.gov/108/plaws/publ364/PLAW-108publ364.pdf.

Brewer, J. (2011). Accessibility of the World Wide Web: Technical and policy perspectives. In W. F. E. Preiser, & K. H. Smith (Eds.), *Universal design handbook* (2nd Edition, pp. 33.1−34.9). New York: McGraw-Hill Companies.

Burgstahler, S. (2017). Universal design of instruction. DO-IT, University of Washington. Retrieved from http://www.washington.edu/doit/Brochures/Programs/ud.html.

Cook, A. M., & Hussey, S. M. (2008). *Assistive technologies: Principles and practice* (3rd Ed). St. Louis, MO: Mosby.

Cook, A. M., & Polgar, J. M. (2015). *Assistive technologies: Principles and practice* (4th Edition). St. Louis, MO: Mosby.

Elliott, T. R., Kurylo, M., & Rivera, P. (2005). Positive growth following acquired physical disability. In C. R. Snyder, & S. J. Lopez (Eds.), *Handbook of positive psychology* (pp. 687−699). New York: Oxford University Press.

Elsaesser, L. J., & Bauer, S. M. (2011). Provision of assistive technology services method (ATSM) according to evidence-based information and knowledge management. *Disability & Rehabilitation: Assistive Technology*, 6(5), 386−401.

Follette Story, M., & Mueller, J. L. (2011). Universal design of products. In W. F. E. Preiser, & K. H. Smith (Eds.), *Universal design handbook* (2nd Edition, pp. 32.1−32.11). New York: McGraw-Hill Companies.

Glanz, B. K., Rimer, F. M., & Lewis, F. M. (2002). The scope of health behavior and health education. In K. Glanz, B. K. Rimer, & F. M. Lewis (Eds.), *Health behavior and health education: Theory, research, and practice* (3rd Edition, pp. 3−21). San Francisco, CA: John Wiley & Sons.

Job Accommodation Network. (2017). Technical assistance manual: Title I of the ADA. Retrieved from https://askjan.org/links/ADAtam1.html#III.

Job Accommodation Network. (2018). Accommodation and compliance series: Providing temporary or trial accommodation solutions. Retrieved from https://askjan.org/topics/temporary.html.

Kennedy, P. (2016). *Inventology: How we dream up things that change the world*. Boston, MA: Houghton Mifflin Harcourt.

Kurtts, S., Dobbins, N., & Takemae, N. (2012). Using assistive technology to meet diverse learner needs. *Library Media Connection*, 30(4), 22−24.

Langton, A. J., & Ramseur, H. (2001). Enhancing employment outcomes through job accommodation and assistive technology resources and services. *Journal of Vocational Rehabilitation*, 16, 27−37.

Lave, J., & Wenger, E. (1991). *Situated learning: Legitimate peripheral participation*. Cambridge, UK: Cambridge University Press.

Meteer, C. A., & Sira, C. S. (2006). Cognitive and emotional consequences of TBI: Intervention strategies for vocational rehabilitation. *NeuroRehabilitation*, 21, 315−326.

Meuller, J. L. (2011). Office and workplace design. In W. F. E. Preiser, & K. H. Smith (Eds.), *Universal design handbook* (2nd Edition, pp. 23.1−23.14). New York: McGraw-Hill Companies.

Nielsen, J. (2017). Usability 101: Introduction to usability. Nielsen Norman Group. Retrieved from http://www.nngroup.com/articles/usability-101-introduction-to-usability/.

Öberg, M. (2008). Approaches to audiological rehabilitation with hearing aids: Studies on pre-fitting strategies and assessment of outcomes. Academic Archive Online. Retrieved from http://www.diva-portal.org/smash/record.jsf?pid = diva2:404.

Ostroff, E. (2001). Universal design: An evolving paradigm. In W. F. E. Preiser, & K. H. Smith (Eds.), *Universal design handbook* (2nd Edition, pp. 1.3−1.11). New York: The McGraw-Hill Companies, Inc.

Ryan, R. M., & Deci, E. L. (2000). Self-determination theory and the facilitation of intrinsic motivation, social development, and well-being. *American Psychologist, 55*(1), 68.

Schwanke, T. D., & Smith, R. O. (2005). Assistive technology outcomes in work settings. *Work: A Journal of Prevention, Assessment and Rehabilitation, 24*(2), 195–204.

Smith, R.O. (2011, September). Universal design & disability [PowerPoint Slides]. Retrieved from https://uwm.courses.wisconsin.edu/d2l/le/content/425320/Home.

Smith, R. O., & Scherer, M. J. (1998). Where are we headed with assistive technology outcomes? In L. U. Vitalini (Ed.), *Volume I: RESNA resource guide for assistive technology outcomes: Measurement tools* (pp. 1–79). RESNA.

United States Equal Employment Opportunity Commission. (2018). Enforcement guidance: Reasonable accommodation and undue hardship under the Americans with Disabilities Act. Retrieved from https://www.eeoc.gov/policy/docs/accommodation.html.

Wehman, P., Inge, K. J., Revell, W. G., & Brooke, V. A. (2007). *Real work for real pay: Inclusive employment for people with disabilities.* Baltimore, MD: Paul H. Brookes Publishing Co.

Accommodation System: Situated experience

5

Anthony Shay

Capacity Building Specialist, Assistive Technologist, and Rehabilitation Specialist, University of Wisconsin-Stout Vocational Rehabilitation Institute (SVRI), Menomonie, WI, United States

THE ACCOMMODATION SYSTEM: SITUATED EXPERIENCE

INTRODUCTION

The motivation to develop a sense of self is present regardless of disability. Determining a vocational goal begins with making choices and then acting on them: we set goals, make intermediate adjustments, and attain the symbols that are indicative of our work toward these goals. We build an identity as a matter of engagement toward, within, and as focused on future endeavors (related to an initial identity goal or otherwise). The identity trajectory upon which we embark is punctuated with the markers of learning and development (Wenger-Trayner & Wenger-Trayner, 2015).

Progress along this journey informs the process of self-definition. Goal striving waxes and wanes. Goals and the importance we assign them may change as we move along our lifelong identity trajectory (Gollwitzer & Kirchhof, 1998). Situated experiences, then, are more than simply a product of the tasks and activities in which we engage while on our employment journey. Situated experiences stand at the nexus of the domains of self, others, environment, and accommodations. The centrality of the self within the accommodation system (AS) model is due the fundamental importance of the self-observing I to understanding and appreciating these experiences. They manifest in task engagement and are characterized by an individual's self-efficacy (vocational competence) (Bandura, 1977; Deci & Moller, 2005), stimulation (level of arousal) (Lewin, 1999), affective state (Csikszentmihalyi, 1990; Feldman Barrett & Russell, 2015), motivational quality of the experience, and the satisfaction and meaning derived from the experience (Csikszentmihalyi, 1990). It is the act of striving to achieve our goals and the dynamic and motivational aspects that lead to our overall valuation of these experiences.

JOB DEVELOPMENT: IDENTITY GOAL STRIVING

Individuals with disabilities seeking to enter or reenter the world of work develop a vocational goal requiring self-definition. The work of gathering the symbols and indicators of the sought after vocational goal sets us upon our identity trajectory (Wicklund & Gollwitzer, 1982). We seek not only to align and engage with established high-status identity relevant personalities, to wear the employment-related "badges of their being" (Agee & Evans, 1966, p. x), but to chart a personal course through which we differentiate and define ourselves within identity relevant contexts (Csikszentmihalyi, 1990; Luft, 1969). We identify with some indicators and disidentify with others (Wenger-Trayner & Wenger-Trayner, 2015). Identity trajectories define the boundaries that give structure to goal striving and make clear where spans (e.g., education, accommodations) are needed to bridge gaps where we find them. The types of goals we set are not all the same. Identity goal striving is different from simple goal-related activity. Simple goals can be said to have been achieved based on a single criterion. A goal of driving to the store is attained once we arrive at the store. A goal of getting a cup of coffee is attained once we have the coffee mug in our hands. Identity goal striving is measured by multiple means. An employment goal of being a nurse requires more than simply donning a nursing outfit. It involves obtaining the symbols of nursing (i.e., wearing the uniform, carrying a stethoscope, reading patient charts, etc.). But it also requires several other indicators such as a formal education, practicums, official licensure, and recognition by a certifying body, among others.

Identity development begins very early in life (Lewin, 1999). It paints the mosaic of personality with vivid colors. As we mature and personality factors influence our vocational choices, we begin a process of gathering the abstract and material symbols relevant to defining the vocational choices we make. Identity indicators progress from those readily attainable to those more difficult to achieve. Humans are social creatures. Identity-building is inexorably intertwined with the people who populate our lives (Luft, 1969). We seek affirmation from others regarding our efforts toward goal attainment. As we determine an employment goal, the possession of employment symbols or indicators becomes increasingly more important. We ascribe even more significance to those we associate with individuals with high status within a desired vocation. Symbols, or identity indicators, are obtained as a means to "stir up readiness in the community to respond to the symbol. And the type of response is not a complicated one. The symbol is effective as long as it causes the community to acknowledge the person's self-definition" (Wicklund & Gollwitzer, 1982, p. 5). An aspiring nurse may begin developing this identity goal by overt admiration of the profession, seeking praise in high school for having good grades in biology, collecting texts related to nursing, taking a nursing assistant or similar job in a nursing home or hospital (proximity to the profession), entering a college program, earning a degree in nursing, and eventually working in a nursing capacity (Lave & Wenger, 1991).

The identity development process may be impeded or halted at any point. Setbacks are to be expected as a natural part of learning and development; however, functional limitations which are not effectively accommodated can introduce negativity and obstacles to developing vocational goals or to the perception of their attainability. Discouragement of this kind in a burgeoning identity whether in a youth preparing to leave secondary school or in an injured worker preparing to return to work can result in a highly negative personal experience leading to poor and ineffectual goal seeking behavior or an outright end to identity striving (i.e., identity incompleteness) (Gollwitzer & Kirchhof, 1998). Individuals must believe they can deal with threats to identity development (Gollwitzer, Marquardt, Scherer, & Fujita, 2013) and that they have the capacity to overcome adversity (Duckworth, 2016). We may substitute a related goal in the short-term, but these are not long-term solutions. Self-symbolizing requires social validation and access to core self-defining elements (Wicklund & Gollwitzer, 1982). In other words, we may seek a vocational identity as a professional football player, but a broken leg challenges the self-defining elements of physical practice. A substitute goal may be more intensive study of the team playbook until our leg heals. We may find, following a little time to heal, that assistive technology such as a leg brace allows ready engagement in self-defining through physical activity. However, impatience is a hallmark of self-symbolizing. Effectively building task efficacy requires of us to be aware of the possibilities which may impact task engagement (Erikson, 1980). Pushing too hard and too fast to obtain identity indicators can jeopardize progress toward our employment goals within the social domain in many ways:

- Impeding social acceptance, and receiving validation from others;
- Creating barriers to building and maintaining meaningful relationships;
- Shutting down effective communication;
- Having a negative effect on our psychological and physical well-being (e.g., we may find the expertise of others demoralizing rather than inspirational) (Langer, 2005; Wicklund & Gollwitzer, 1982).

As we build an identity and develop our own expertise, the need for continued validation from others is reduced (Wicklund & Gollwitzer, 1982).

COMPENSATORY STRATEGIES TO COMPENSATORY ACTION

Compensatory action is the renewed and redoubled effort an individual invests in self-definition following an end to threatened or halted identity striving. When the limitations imposed by disability are removed or reduced, identity (employment goal) striving resumes in earnest. Our central focus becomes successful identity goal achievement. Professionals working with job seekers who have disabilities may harness this compensatory action toward more effective goal striving (Matschke, Fehr, & Sassenberg, 2012). They are in a position to anticipate an identity trajectory and be the first to affirm, facilitate, or reinforce identity indicator selection for those with whom they are working. Likewise, the idea that accommodations, such as assistive technology, are instrumental to identity completion is an important piece of information that must be shared and integrated into employment plan development. Effective assessment toward the provision of effective accommodations is of critical importance (Shay, Anderson, & Matthews, 2017).

Poorly matched, ineffective, and/or faulty assistive technology (or other accommodations) may have the opposite effect of a compensatory action and may serve to not only halt progress but take an individual to a developmental point prior to the original accommodation point—reinforcing the impediment to employment rather than the compensatory effort. As disability-employment professionals, we need to emphasize the role accommodations play in identity striving behavior. Framing accommodation use (and user attitudes) is a significant component in the development and maintenance of identity goal striving behavior (Shay et al., 2017). Individuals with disabilities should not only reflect on the role accommodations play in identity goal achievement, professionals working with them need to be aware of and capitalize upon the compensatory goal reinvigoration as a springboard to renewed vocational identity striving efforts. When we optimize work and acknowledge the satisfaction and meaning we find in it—the role and importance of accommodations in job maintenance becomes a necessary point of discussion. Ambiguous feelings related to accommodation use (e.g., stemming from stigma, technology tolerance, self-consciousness) may be mitigated (De Jong, Scherer, & Rodger, 2007); we may find greater value in task engagement than accommodation abandonment. Accommodation use and identity goal striving are ever changing and lifelong processes. The relationship between the gathering of identity symbols as they relate to vocational goal striving and the ability to accommodate this activity is critical to obtaining and maintaining employment and should not be underestimated or overlooked (Shay et al., 2017).

GOAL STRIVING AND ADAPTIVE BEHAVIOR

Work contexts are dynamic. So too is human development. Learning occurs over time and practice. Adaptive work behavior is a product of stability and reinforces positive patterns of task engagement. Optimizing work behavior requires believing that people with disabilities can develop beyond the level at which employment and disability interventions occur (Feuerstein, Falik, & Feuerstein, 2015). Bridging the developmental gap requires effort and persistence over time (Duckworth, Peterson, Matthews, Kelly, & Carver, 2007; Vygotsky, 1978). Motivation is optimized when intermediate goals are established and modified to ensure there is an equilibrium between the challenges experienced in a job task and the skills and abilities we have to meet these demands. In-task negotiations such as this are imbued with resiliency and stimulate motivation, necessary components when dealing with both voluntary and involuntary task-engagement factors (Ajzen, 1991; Csikszentmihalyi, 1990). Indeed, developing a pattern of adaptive behavior not only increases psychological complexity (Csikszentmihalyi, 1990) but also facilitates job maintenance (Dawis & Lofquist, 1984).

A SITUATED EXPERIENCE MODEL

The situated experience model in Fig. 5.1 illustrates several zones, or ambits of influence, which delineate the quality of personal experience we can expect given the balance we are able to achieve (i.e., what we experience and how intense the experience is). With each successive time we engage in a task, we seek to maximize stimulation and positive affect by achieving a higher than average balance between the challenges we face and the skills with which we meet them. This situated experience influences motivation and quality of work life:

- Low-task challenge + low-task efficacy = an ambit of apathy. This ambit is characterized by negative emotions and low stimulation.
- Low-task challenge + high-task efficacy = an ambit of complacency. This ambit is characterized by positive emotions and low stimulation.
- High-task challenge + high-task efficacy = an ambit of optimal experience. This ambit is characterized by positive emotions and high stimulation.
- High-task challenge + low-task efficacy = an ambit of anxiety. This ambit is characterized by negative emotions and high stimulation.

The model illustrates the quality of our personal experience within the work tasks we perform while on the job. As noted above, situated experience considers

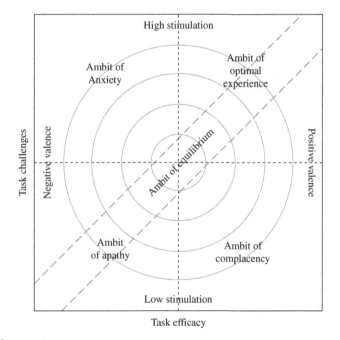

FIGURE 5.1

Situated experience model.

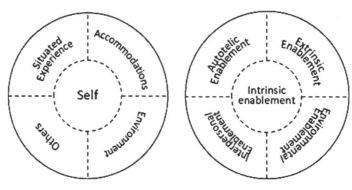

FIGURE 5.2

The accommodation system and enablement domains.

- How balanced a work task is in terms of the challenges we face, and how effectively our skills match them (i.e., our vocational competence);
- How much stimulation there is within a task;
- How we feel when engaged in a work task;
- The motivational quality of the experience;
- The satisfaction and meaning we derived from the overall experience (Fig. 5.2).

The model provides us with an overall frame by which to assess and discuss personal experiences on a job (or in any task or activity). Through mapping, scaffolding, and optimizing task engagement and goal striving behavior, we can more effectively provide positive, rewarding, and satisfying experiences for people with functional limitations regardless of ability. Table 5.1 highlights some characteristics of negative and positive situated experiences across the AS domains of self, others, accommodations, and environment (items listed are not mutually exclusive or exhaustive).

ADAPTIVE WORK BEHAVIOR

We prioritize our behavior to meet our needs. As situations change and as our needs are met, our valuation of these needs change. The degree to which our needs are met creates a psychological state of tension we must negotiate in our ongoing efforts to reach our goals (Lewin, 1999). We have the capacity to make these evaluations, but there is an inverse relationship between the quantity and the quality we can achieve. We can either generate many ideas of low quality or we can come up with a few high-quality ideas—as one element increases the other decreases. How we make the evaluations is a subjective process (Perkins, 1981, p. 142). As we develop and learn to negotiate situated experiences,

Table 5.1 Characteristics of Positive and Negative Situated Experiences

Negative Situated Experience	Positive Situated Experience
Self: Intrinsic Enablement	
Poor coping skills	Effective coping skills / behavior is adaptive
Demonstrates maladaptive behavior	Responsible, dependable, reliable
Poor attitude	Perceived behavioral control
Stigmatization	Balanced task engagement stimulation
Lack of soft skills	Task optimization behavior
Easily distracted	Task efficacy
Multiple disabilities / functional impairments / frailty which are inadequately addressed	Has an established vocational identity (reflects the Great 8)
Situated Experience: Autotelic Enablement	
Poor balance between challenges and skills	Demonstrates optimization behavior
Self-consciousness	Openness to and immediacy of feedback
Poorly defined goals	Able to set intermediate goals
Identity incompleteness	Reengagement in goal-directed behavior
Excessive identity indicator substitution	Compensatory action
Choosing extrinsic reasons for task engagement (e.g. money, benefits, recognition)	Choosing intrinsic reasons for task engagement (e.g. enjoyment, wellbeing, personal enrichment)
Accommodations: Extrinsic Enablement	
Accommodation abandonment	Accommodation goodness of fit
Poor fit between technology and needs	Well-matched assistive technology features
Environmental barriers	Availability/access to compensatory
Lack of accommodation-related training for self or others (e.g. job coach)	strategies
	Consumer informed choice / preferences considered
Environment: Environmental Enablement	
Person is expected to adapt to the Environment	Universal design principles utilized
Lack of resources	Access to necessary resources
Impoverished environments	Access to the built environment
Lack of necessary resources	Environments optimized for challenge / skill
Legislation and policies reinforcing stigma and discrimination	balance
	Legislation and policies supporting service needs of individuals
Others: Interpersonal Enablement	
Poor/lack of communication	Effective communication
Social isolation/marginalization	Access to identity indicators/symbols
Passive social engagement	Access to expertise and training
Lack of identity affirmation	Validation of vocational identity
Socioeconomic barriers	Active social engagement

psychological complexity and competence increases (Csikszentmihalyi, Abuhamdeh, & Nakamura, 2005; Lewin, 1997).

The role of employment professionals providing services to people with disabilities centers on facilitating effective work behavior and environments conducive to meaningful and satisfying work:

- Knowing and understanding what is expected;
- Receiving immediate feedback during work activity;

- Being completely focused on job task requirements;
- Avoiding distractions on the job;
- Having a sense of control within the activity;
- Losing our sense of self while engaged in a job task (a sense of merging between self and the task);
- A lack of boredom while engaged in the work we are doing;
- A lack of effortful exertion to maintain focus on what we are doing (effort comes more naturally and loses the "work" intentionality);
- A tendency toward task engagement for the sake of the engagement (rather than only for external rewards such as pay, recognition, or benefits);
- A distortion of our sense of time. While we are focused on the work we are doing, time seems to speed up or slow down;
- We tend to enjoy the work activity (Csikszentmihalyi, 1975, 1990).

Finding satisfying work does not hinge exclusively on fulfilling all of these criteria (Shay, 2009). Work that reflects most or all of these qualities tends to be highly self-reinforcing—our motivation is interiorized and continuing the behavior is self-reinforced. Being mindful within an activity helps keep us centered or focused on what we are doing. Ellen Langer describes this as a creative process whereby we more easily recognize a state of disequilibrium and work back toward balance—a practice leading to a "personal renaissance" (Langer, 2005, p. 20). Langer's concept is similar to Csikszentmihalyi's (1990) concept of self-reinforced behavioral motivation as this develops over time so too does our psychological complexity. Any type of work activity can be reflective of self-reinforced behavioral motivation. People who tend to seek out highly rewarding activities will seek to optimize the work within which they engage. They choose activities that are challenging and which will utilize their skills at a high level (Csikszentmihalyi, 1990; Csikszentmihalyi et al., 2005; Shay, 2009). Our personal experience of the balance between challenges faced and the skills we employ within work tasks (i.e., how well we are able to negotiate task requirements within specific work contexts—our situated experiences) dictates our attitudes, expectations, beliefs, and continued intention to engage in the specific work tasks.

OPTIMIZING TASK BEHAVIOR

An ambit of equilibrium may exist in situated experiences where there may not necessarily be high task challenges and a high level of task-efficacy (Csikszentmihalyi, 1975). The overarching goal would be to experience a higher than average balance between challenges and skills as a means to optimize motivation, positive task regard, and to build or reinforce psychological complexity (as reflected by the concentric circles in Fig. 5.1) (Csikszentmihalyi, 1990). People engaging in novel tasks or those who may be unable to appreciate practice history or maintain task awareness can still achieve positive, satisfying, and rewarding vocational experiences through a task—challenge equilibrium. They may not be able to achieve a higher than average experience but can still have their extant skill levels matched by established (e.g., modified) task challenges. Furthermore, optimal experiences may instill a beneficial work orientation. This can also guard against more negative outcomes into the future (Csikszentmihalyi et al., 2005; Feuerstein et al., 2015). To optimize behavior and self-reinforce our motivation for a task we seek a continual (stepwise) increase in challenges as we increase our skills. We strive to maintain task challenge that is slightly above our skill level since time on task (i.e., effort and persistence over time) gradually moves us in the direction of skill development (Csikszentmihalyi, 1990). Goal striving behavior is optimized when we seek to increase skill or challenge levels within a task or by switching to a different task altogether. We optimize this balance when we:

> Increase skills: Attempt to increase our skill level in a work task when task challenges increase;
> Increase skills: Change to another work task that we have the skills to perform when we determine we do not have the necessary skills for the work task;
> Increase challenges: Attempt to increase the challenge in a work task when we find them too easy;
> Increase challenges: Change to another work task which requires greater skills to perform when we determine that our skills exceed that required by the task in which we are curently engaged (Csikszentmihalyi, 1990).

Effective identity development involves working toward stability (Erikson, 1980). As stated above, we set and modify goals to maintain a balance within work tasks. Work-related changes and engaging in new and unfamiliar tasks upsets this balance (e.g., changes related to the environment, ourselves, others, accommodations). Job stabilization requires us to reestablish balance (Csikszentmihalyi, 1990; Csikszentmihalyi et al., 2005).

AN AMBIT OF IMBALANCE

Task competence is the realization of the balance we experience between challenges faced on the job and our ability to meet them. Task attempts leading to excessive challenges and a lack of necessary skills leads to feelings of anxiety, overstimulation, and little motivation for continued task engagement (Allison & Carlisle Duncan, 1988; Csikszentmihalyi, 1990; Csikszentmihalyi et al., 2005; Lewin, 1999; Massimini, Csikszentmihalyi, & Carli, 1987). We perceive behavioral control as being unlikely. Likewise, a high level of skills met with a low level of task challenges leads to feelings of indifference or apathy toward a job, understimulation, and low motivation for continued task engagement. Negativity pervades our subjective norms and attitudes in both of these scenarios (Ajzen, 1991).

THE PERCEPTION OF CONTROL

Our intention toward task engagement (i.e., our motivation) and the belief in our ability to achieve the goals, we set for ourselves the stage for increased quality of work life and work satisfaction—the precursor conditions for goal striving. Attitudes, subjective norms (which includes moral dispositions), and perceived behavioral control guide our intention for initial and sustained goal striving behavior. The shorter the time interval between our intention to engage in a task and actual engagement as well as the shorter the difference between our perceived and actual behavioral control, the more confident we can be in the likelihood of continued task engagement (Ajzen, 1991). We can define rewarding and meaningful work, then, as intentional work behavior sustained over time leading to the expectation of a sense of control over our work activity. This is a personal experience reflective of task challenges and skills balanced at a high level, the belief that this can be sustained on an ongoing basis and which leads to positive personal beliefs, attitudes, and feelings toward work task engagement (i.e., a positive situated experience) (Ajzen, 1991; Csikszentmihalyi, 1990). These positive situated experiences drive job maintenance. Fig. 5.3 illustrates a case study map for Erik regarding elements of his situated experience.

SUMMARY

Disability represents a challenge to developing an identity but also provides an opportunity through matching appropriate accommodations to reengage and reinvigorate the goal striving process. A buoyed identity can be kept afloat through maintaining and optimizing the task challenges with a commensurate level of task efficacy. A person-centered process based on consumer informed choice drives goal setting, striving, and achievement leading to quality of work life and job satisfaction over the long term. Identity trajectories are unique and lifelong (Wenger-Trayner & Wenger-Trayner, 2015). Our situated experiences color our perceptions, beliefs, and attitudes toward goal striving and working to complete the vocational identities we select for ourselves. Social validation cements identity completeness and reinforces striving as we age albeit on a diminishing basis (Gollwitzer & Kirchhof, 1998). When working with individuals with disabilities seeking employment, approaching service delivery from the standpoint of facilitating identity development holds a key to framing the process in terms of the role and importance of accommodation use to identity goal development, maintenance, and achievement and the ramifications they have on job satisfaction and quality of work life.

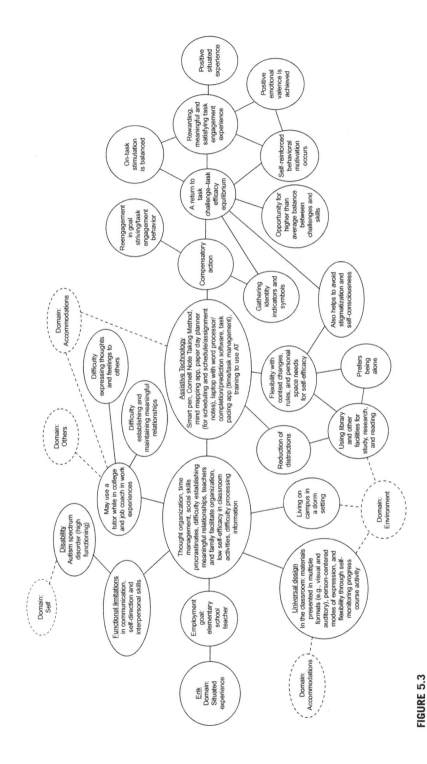

FIGURE 5.3

Erik case study map: situated experience.

REFERENCES

Agee, J., & Evans, W. (1966). *Many are called*. Boston: Houghton Mifflin.

Ajzen, I. (1991). The theory of planned behavior. *Organizational Behavior and Human Decision Processes, 50*, 179−211.

Allison, M. T., & Carlisle Duncan, M. (1988). Women, work, and flow. In M. Csikszentmihalyi, & I. S. Csikszentmihalyi (Eds.), *Optimal experience: Psychological studies of flow in consciousness* (pp. 118−137). New York, NY: Cambridge University Press.

Bandura, A. (1977). Self XE "Self"-efficacy: Toward a unifying theory of behavioral change. *Psychological Review, 84*(2), 191.

Csikszentmihalyi, M. (1975). *Beyond boredom and anxiety*. San Francisco, CA: Jossey-Bass Inc., Publishers.

Csikszentmihalyi, M. (1990). *Flow: The psychology of optimal experience*. New York, NY: Harper Perennial.

Csikszentmihalyi, M., Abuhamdeh, S., & Nakamura, J. (2005). Flow. In A. J. Elliot, & C. S. Dweck (Eds.), *Handbook of competence and motivation* (pp. 598−608). New York, NY: The Guilford Press.

Dawis, R. V., & Lofquist, L. H. (1984). *A psychological theory of work adjustment*. Minneapolis, MN: University of Minnesota Press.

Deci, E. L., & Moller, A. C. (2005). The concept of competence. In A. J. Elliot, & C. S. Dweck (Eds.), *Handbook of competence and motivation* (pp. 579−597). New York, NY: The Guilford Press.

De Jong, D., Scherer, M. J., & Rodger, S. (2007). *Assistive technology in the workplace*. St. Louis, MO: Mosby.

Duckworth, A. (2016). *Grit: The power of passion and perseverance*. New York, NY: Scribner.

Duckworth, A., Peterson, C., Matthews, M., Kelly, D., & Carver, C. S. (2007). Grit: Perseverance and passion for long-term goals. *Journal of Personality and Social Psychology, 92*(6), 1087−1101.

Erikson, E. H. (1980). *Identity and the life cycle*. New York, NY: W. W. Norton & Company, Inc.

Feldman Barrett, L., & Russell, J. A. (2015). An introduction to psychological construction. In L. Feldman Barrett, & J. A. Russell (Eds.), *The Psychological Construction of Emotion* (pp. 1−17). New York, NY: The Guilford Press.

Feuerstein, R., Falik, L. H., & Feuerstein, R. S. (2015). *Changing minds & brains: The legacy of Reuven Feuerstein*. New York, NY: Teacher's College Press.

Gollwitzer, P. M., & Kirchhof, O. (1998). The willful pursuit of identity. In J. Heckhausen, & C. Dweck (Eds.), *Motivation and self-regulation across the lifespan* (pp. 389−423). Cambridge, New York: Cambridge University Press.

Gollwitzer, P. M., Marquardt, M. K., Scherer, M., & Fujita, K. (2013). Identity-goal threats engaging in distinct compensatory efforts. *Social Psychological and Personality Science, 4*(5), 555−562.

Langer, E. J. (2005). *On becoming an artist: Reinventing yourself through mindful creativity*. New York, NY: Ballantine Books.

Lave, J., & Wenger, E. (1991). *Situated learning: Legitimate peripheral participation*. Cambridge, UK: Cambridge University Press.

Lewin, K. (1997). Conduct, knowledge, and acceptance of new values. In G. W. Lewin (Ed.), *Resolving social conflicts & field theory in social science* (pp. 49–55). Washington, DC: American Psychological Association.

Lewin, K. (1999). Level of aspiration. In M. Gold (Ed.), *The complete social scientist* (pp. 137–182). Washington, DC: American Psychological Association.

Luft, J. (1969). *Of human interaction: The Johari model.* Palo Alto, CA: Mayfield Publishing Company.

Massimini, F., Csikszentmihalyi, M., & Carli, M. (1987). The monitoring of optimal experience: A tool for psychiatric rehabilitation. *The Journal of Nervous and Mental Disease*, *175*(9), 545–549.

Matschke, C., Fehr, J., & Sassenberg, K. (2012). When does goal discrepancy induce compensatory effort? An application of self-completion theory to social issues. *Social and Personality Psychology Compass*, *6*(7), 536–550.

Perkins, D. N. (1981). *The mind's best work.* Cambridge, MA: Harvard University Press.

Shay, A.F. (2009). *An exploratory study of optimal experience in individuals with limitations in attention.* (Doctoral dissertation). Available from ProQuest LLC.

Shay, A. F., Anderson, C. A., & Matthews, P. (2017). Empowering youth self-definition and identity through assistive technology assessment. *Vocational Evaluation and Work Adjustment Association Journal*, *41*(2), 78–88.

Vygotsky, L. S. (1978). Interaction between learning and development. In M. Cole, V. John-Steiner, S. Scribner, & E. Souberman (Eds.), Mind in society*: The development of higher psychological processes* (pp. 79–91). Cambridge, MA: Harvard University Press.

Wenger-Trayner, E., & Wenger-Trayner, B. (2015). Learning in a landscape of practice: A framework. In E. Wenger-Trayner, M. Fenton-O'Creevy, S. Hutchinson, C. Kubiak, & B. Wenger-Trayner (Eds.), *Learning in landscapes of practice: Boundaries, identity, and knowledgeability in practice-based learning* (pp. 13–29). New York, NY: Routledge.

Wicklund, R. A., & Gollwitzer, P. M. (1982). *Symbolic self-completion.* New York, NY: Routledge.

The assistive technology service delivery process

2

ASSISTIVE TECHNOLOGY SERVICE DELIVERY IN THE CONTEXT OF EMPLOYMENT

In the disability-employment service delivery (DESD) process, consumers move through stages much like those in the assistive technology service delivery (ATSD) process. In Part 2 of this text, we frame the ATSD process in much the same way disability-employment services are delivered (Langton, Hughes, Flynn, Gaster, & Augustine, 1994)—providing a readily accessible framework to professionals in both fields. Fig. P2.1 illustrates the employment analog in relation to the ATSD and DESD processes. Colling and Davis in their article *The counseling function in vocational rehabilitation* provide elements helpful in delineating the person-centered focus of these processes. The elements are not mutually exclusive:

- Attending: Information gathering focused on the nature of a consumer's needs in relation to employment goal striving, building a working alliance, and setting expectations for moving forward toward gainful employment;
- Exploration: Information gathering continues, the consumer's Great 8 are considered (Chapter 1: Accommodation System: Self), and employment goals are considered with preliminary consideration of goals and necessary services toward goal achievement.
- Understanding: Goals and progress measures are developed, services, and potential barriers are determined, and an action plan is developed which is designed to help the consumer achieve their employment goal;
- Action: The consumer engages in employment-related services which further clarifies needs, barriers, and next steps in conjunction with the service professional's continued guidance and support (i.e., understanding) toward achieving their established employment goal;
- Termination: Case closure is framed as a positive end to productive service delivery. It is foreshadowed throughout service delivery to prepare consumers for the eventual end of programmatic involvement and transition to the next steps in the disability-employment service process (Colling & Davis, 2005).

Through the services we provide, we facilitate community inclusion (Dutta, Shiro-Geist, & Kundu, 2009) with the overarching goal of consumer

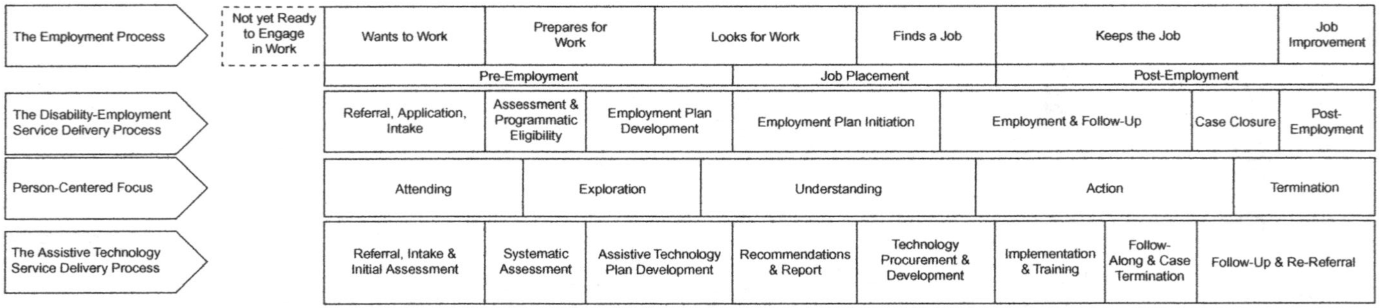

The Employment Process	Wants to Work	Prepares for Work	Looks for Work	Finds a Job	Keeps the Job	Job Improvement
(Not yet Ready to Engage in Work)						
The Disability-Employment Service Delivery Process	Pre-Employment				Job Placement / Post-Employment	
	Referral, Application, Intake	Assessment & Programmatic Eligibility / Employment Plan Development	Employment Plan Initiation		Employment & Follow-Up	Case Closure / Post-Employment
Person-Centered Focus	Attending	Exploration	Understanding		Action	Termination
The Assistive Technology Service Delivery Process	Referral, Intake & Initial Assessment	Systematic Assessment / Assistive Technology Plan Development	Recommendations & Report	Technology Procurement & Development	Implementation & Training / Follow-Along & Case Termination	Follow-Up & Re-Referral

FIGURE P2.1

A comparison of the employment, assistive technology service delivery, and disability-employment processes.

empowerment. We seek not only to develop consumer self-reliance but also the development of effective behavior. To this end, service delivery provides structure, consistency, role modeling, and the mediation of task meaning with and for consumers as appropriate (Sampson, Reardon, Peterson, & Lenz, 2004; Feuerstein, Feuerstein, & Falik, 2010). Consumer's learn through engagement in service delivery processes how to more independently engage in these same activities—meaningful disability-employment activities—in the future.

In Fig. P2.1, we offer a side-by-side comparison of processes; however, it is important to note that a consumer may begin the disability-employment process already having found (or attempting to maintain) employment. Assistive technology services may begin at any point within DESD (e.g., to allow the consumer the ability to access services or following case closure in postemployment to address job retention concerns).

THE ASSISTIVE TECHNOLOGY SERVICE DELIVERY PROCESS

The six chapters which comprise Part 2 of this text offer an overview of the ATSD process over five chapters. Chapter 6, Overview of the Service Delivery Process offers a general overview of the ATSD process from a global perspective. Chapter 7, Referral, Intake, and Assessment considers the elements of service entry: referral, intake, and assessment. Chapter 8, Plan Development, Recommendation, and Report addresses the mid-process components of plan development, recommendations, and report. Chapter 10, Technology Procurement and Development provides a perspective on technology procurement and development and Chapter 11, Implementation and Training focuses on the implementation and training of assistive technologies. Chapter 12, Follow-Up, Follow Along, Service Completion, and Outcomes considers service exit in the follow-up, follow-along, and service completion stages of the service delivery process. A chapter on problem-solving (Chapter 9: Problem Solving) precedes technology procurement and development since this is a critical aspect in arriving at a person-centered (i.e., technology user-centered) solution—which should reduce the incidence of technology abandonment (discussed in Chapter 15: On Technology Abandonment or Discontinuance). The end of the ATSD process—follow-up, follow along, and service completion—is followed by a discussion of assistive technology outcomes (Chapter 12: Follow-Up, Follow Along, Service Completion, and Outcomes) which are important in determining and reporting program effectiveness, consumer satisfaction, and that assistive technologies being implemented are efficacious in meeting consumer needs.

REFERENCES

Colling, K., & Davis, A. (2005). The counseling function in vocational rehabilitation. *Journal of Applied Rehabilitation Counseling, 36*(1), 6–11.

Dutta, A., Shiro-Geist, C., & Kundu, M. (2009). Coordination of postsecondary transition services for students with disabilities. *Journal of Rehabilitation, 75*(1), 10–17.

Feuerstein, R., Feuerstein, R. S., & Falik, L. H. (2010). *Beyond smarter: Mediated learning and the brain's capacity for change.* New York, NY: Teacher's College Press.

Langton, A. J., Hughes, J. L., Flynn, C. C., Gaster, L. S., & Augustine, V. (1994). *Tech points: Integrating rehabilitation technology into vocational rehabilitation services training manual.* West Columbia, SC: South Carolina State Vocational Rehabilitation Department, Center for Rehabilitation Technology. Retrieved from https://files.eric.ed.gov/fulltext/ED405688.pdf.

Sampson, J. P., Reardon, R. C., Peterson, G. W., & Lenz, J. G. (2004). *Career counseling and services: A cognitive information processing approach.* Belmont, CA: Brooks/Cole.

Overview of the assistive technology service delivery process: An international perspective

6

Marcia Scherer

*Institute for Matching Person and Technology, University of Rochester Medical Center;
Physical Medicine and Rehabilitation and Senior Research Associate, International Center for
Hearing and Speech Research, Webster, NY, United States*

THE ASSISTIVE TECHNOLOGY SYSTEM

Assistive technology (AT) service delivery (ATSD) takes place within an AT system. The components of this system include users and their families, AT products, AT services, personnel, service providing agencies, manufacturers, distributers, funding agencies, and policies and legislation. The components of the AT systems vary substantially within and across countries, as do the interactions between the components. For example, some manufacturers work closely with users on the design of assistive products while others do not; in some countries, a broad range of products and services are available, while in others they are limited; and in some contexts, the national or local government carry the cost for AT, in other contexts, users and their family members need to pay fully for them, and in yet other contexts, they are paid for through donations.

ASSISTIVE TECHNOLOGY SERVICE DELIVERY

In many countries, the delivery of AT services follows a structure similar to the three-level healthcare system. Basic needs for AT are met at primary (municipality/community) level, while more advanced needs are met at secondary (county) level. At instances where secondary level services cannot meet the needs, they may call on tertiary (national) level services. According to the Norwegian AT System Model, AT Centers in each county work closely with healthcare and rehabilitation service providers at the municipal level. Service delivery spans a continuum whereby service provision moves from nationally focused, infrequent, complicated, and specialized to locally focused, frequent, simple, and more generalized services. See Fig. 6.1 for an illustration of the Norwegian Assistive Technology System Model.

Assistive Technology Service Delivery. DOI: https://doi.org/10.1016/B978-0-12-812979-1.00006-0

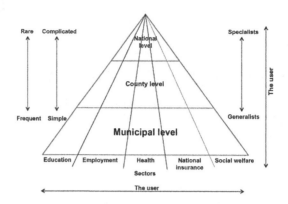

FIGURE 6.1

Norwegian Assistive Technology System Model.

In Europe, the Horizontal European Activities in Rehabilitation Technology (HEART) study was conducted from 1993 to 1995 to survey AT services in Europe and to propose positive future actions (Horizontal European Activities in Rehabilitation Technology Consortium, 1995). Led by the Swedish Institute of Assistive Technology, the study team consisted of 21 institutions, organizations, and companies in 12 European countries. The HEART study defined six criteria for good service delivery systems:

1. **Accessibility** to those who need it. Accessibility is based on the needs of the individual and it means people know where to go for help, that the system is easy to understand, and that ample information about it is available. No one should be excluded regardless of where they live or their ability to pay. These points were underscored in the 2007 report on the Future of Disability in America (Field & Jette, 2007). We can conclude, therefore, that many of the HEART criteria have both current and international applicability.
2. **Competence** of the people working in the service delivery system. This means that personnel have the knowledge, skills, and experience necessary to provide proper service to the users of the system.
3. **Coordination of the parts** into a cohesive whole
   a. between the individual and all steps in the delivery process
   b. between the professionals working with service delivery
   c. between the service delivery system and other sectors of society.
4. **Efficiency** in finding the best solutions for the greatest number of people, using available resources at the lowest cost in the shortest possible time.
5. **Flexibility** in responding to change and allowing for individual differences. Flexibility is vital for AT because of the diverse nature of disabilities and constant technological advances. Flexibility exists when
   a. potential users can get devices that meet their needs whether or not those devices are on an approved list or are marketed in the countries where users live

   b. producers—importers can get a device tested at a reasonable cost, within a reasonable time
   c. researchers and developers can get financial support for their work; cooperate and communicate with users, designers, and producers; and utilize new technology to meet needs.
6. **User influence** as indicated by
   a. the presence and strength of user organizations
   b. the enforcement of user legal rights
   c. user involvement on the policy-making level
   d. frequent user consultation during the process
   e. user influence in the decision-making process.

VARIOUS SERVICE DELIVERY PROCESSES

The processes of delivering AT services vary between and within countries. A study of the ATSD processes used in 16 European countries in the 1990s identified the following seven steps:

1. Initiative (the first contact with the service delivery system)
2. Assessment (evaluation of needs)
3. Selection of the assistive solution (defining the individual AT program)
4. Selection of the equipment (choosing the specific equipment within the AT program)
5. Authorization (obtaining funding)
6. Implementation (delivering the equipment to the user, fitting and training)
7. Management and follow-up (maintenance and periodic verification)

In their paper, *Providing Assistive Devices and Rehabilitation Services in Developing Countries*, Øderud and Grann (1999) reflect on experiences from Scandinavia, Europe, and low-and-middle-income countries (LMIC). They outline key elements for the effective provision of assistive devices which emphasize the local availability of devices as well as their repair and maintenance. This requires competent providers and professionals as well as an effective service delivery system. The authors note that it is important to provide assistive technology devices or ATDs to individual users in their local environments so as to ensure a match with the environment as well as the user. Effective use depends on local attitudes as well as local services. For example, some countries formerly under totalitarian regimes have followed the practice of institutionalization and exclusion of individuals with disabilities (Eldar et al., 2008).

In these countries, some professionals, such as social workers, do not exist and individuals do not receive information about available AT, not to mention the devices themselves.

While recognizing local variation, the authors outline a structured service delivery system containing the following core elements (Øderud & Grann, 1999, p. 784):

1. Identifying potential users
2. Identifying user needs
3. Identifying potential ATD solutions
4. Individual matching of user and ATD
5. Training
6. Follow-up with users, personal assistants, caregivers, and families
7. Service and maintenance

Twelve years after Øderud and Grann published their paper, *The World Report on Disability* (World Health Organization, 2011) underscores the same points by highlighting the need to ensure that ATDs are appropriate, suits the environment and the user, and are of high quality with adequate follow-up to ensure safe and efficient use of the ATD. The term *appropriate technology* is used to mean

Assistive technology that meets people's needs, uses local skills, tools, and materials, and is simple, effective, affordable, and acceptable to its users (World Health Organization, 2011, p. 118).

SERVICE STAGES

Information is gathered throughout the service delivery process (although primarily during a formal AT assessment) and is aggregated and evaluated by the AT professional. AT feature matching (i.e., matching specific AT attributes to specific student needs) takes place in collaboration with the student and his/her interdisciplinary rehabilitation team (Cook & Polgar, 2012). A plan is developed with the consumer and a report is provided with AT recommendations identified as an effective fit for the consumer. This is submitted to the referral source/funding agency. Although much information is collected at the front end of the AT assessment, the AT service delivery team remains vigilant in gathering and communicating relevant data throughout the entire process. This assures AT-consumer goodness-of-fit since issues and concerns may creep into the process following the formal assessment (i.e., technical concerns with device customization or fabrication, during setup and implementation, or during follow-along/follow-up) (Shay, Anderson, & Matthews, 2017) (Table 6.1).

Table 6.1 Overview of Key Stages Typically Involved in Assistive Technology Service Delivery

AT Service Delivery Process Steps	Case Management Considerations
1. Referral	Collect referral information and contact consumer; basic information gathered on disability/limitations, referral purpose, assign staff
2. Intake and Initial Assessment	Follow-up with and add to information received during referral, determine service delivery needs, begin gathering information regarding functional skills/limitations, employment goal, begin considering AT plan objectives
3. Systematic Assessment	Complete a formal AT assessment including a comprehensive needs assessment, feature matching to include demonstrations, trial, and simulations as necessary
4. Plan Development	Develop intervention strategies and build in timeframes and progress measures toward meeting consumer informed choice and rehabilitation team expectations
5. Recommendations and Report	Write a review of the assessment, AT plan, recommendations, implementation plan and toward approval to move to next steps
6. Technology Procurement and Development	Ordering or purchasing the AT devises, systems, parts, and materials with customization or fabrication as necessary, toward implementation
7. Implementation and Training	Implementing or installing AT as outlined in the implantation plan including training as needed
8. Follow-Along and Case Termination	Over the short-term measure and document outcomes, reassess as necessary and provide necessary services toward closing the case
9. Follow-Up and Re-referral	Over the longer-term following case closure, measure and document outcomes, reassess as necessary (e.g. need for additional or remedial training) including the need for a new referral to the referral/funding or another source

Adapted from Shay, A.F., Anderson, C.A., Matthews, P. (2017). Empowering youth self-definition and identity through assistive technology assessment. Vocational Evaluation and Work Adjustment Association Journal.

A GLOBAL OUTLOOK ON THE ACCESS TO ASSISTIVE TECHNOLOGY

The 58th World Health Assembly in 2005 addressed disability prevention, management, and rehabilitation (World Health Organization, 2005). The report stated that there were approximately 600 million people globally living with disabilities of various types, and of that total, 80% live in LMIC; most are poor with limited or no access to basic services, including rehabilitation facilities and services.

Limited access to information and services is a problem and, additionally, rehabilitation has not become a part of many healthcare systems worldwide due to many historical, cultural, political, and economic factors (Tinney, Chiodo, Haig, & Wiredu, 2007).

The World Report on Disability documents widespread evidence of barriers to the achievement of this, including the following:

- **Lack of provision of services.** People with disabilities are particularly vulnerable to deficiencies in services such as health care, rehabilitation, or support and assistance.
- **Problems with service delivery.** Issues such as poor coordination among services, inadequate staffing, staff competencies, and training affect the quality and adequacy of services for persons with disabilities.
- **Inadequate funding.** Resources allocated to implementing policies and plans are often inadequate. Strategy papers on poverty reduction, for instance, may mention disability but without considering funding.
- **Lack of accessibility.** Built environments (including public accommodations) transport systems and information are often inaccessible. Lack of access to transport is a frequent reason for a person with a disability being discouraged from seeking work or prevented from accessing health care. Even in countries with laws on accessibility, compliance in public buildings is often very low. The communication needs of people with disabilities are often unmet. Information is frequently unavailable in accessible formats, and some people with disabilities are unable to access basic information and communication technologies such as telephones and television.
- **Lack of consultation and involvement.** Often people with disabilities are excluded from decision-making in matters directly affecting their lives. The WHO states,

  Educating people with disabilities is essential for developing knowledge and skills for self-help, care, management, and decision-making. People with disabilities and their families experience better health and functioning when they are partners in rehabilitation (World Health Organization, 2011, p. 96).

- **Lack of data and evidence.** A lack of rigorous and comparable data on disability and evidence on programs that work often impedes understanding and action (World Health Organization, 2011, pp. 262−263).

The localized approach to service delivery advocated by Øderud and Grann and in *The World Report on Disability* has been successfully used elsewhere for the training of professionals and providers. In the case of ATDs, local rehabilitation workshops emphasize the identification of local materials and supply as well as trained providers and technicians, information dissemination, individualized matching and adaptation of ATDs, and training and follow-up with users and families (e.g., Øderud, Brodtkorb, & Hotchkiss, 2004).

But as noted in *The World Report on Disability*, many barriers remain. While there are globally accessible information resources for ATDs like AbleData (see https://abledata.acl.gov), they often aren't available in the language of individuals in developing countries, and thus, their usability is limited. As an example of an effort to meet a global need for information on a variety of ATDs utilizing low-cost and local materials, the United Nations (1997) developed an excellent resource describing how to use local resources and materials to make prostheses and orthoses. Similarly, Handicap International has developed guidance on how to make and fit wooden wheelchairs, crutches, simple hand grips, stick reachers, swivel and pronged-handles, spoons, special seating and maintain hearing aids.

ENVIRONMENTAL AND PERSONAL FACTORS
THE ATD DECISION-MAKING AND DEVICE SELECTION PROCESS

It is important to assess the consumer's readiness to use a given ATD and to provide the most appropriate device and training tailored to the unique characteristics and issues of that consumer. The ATD provider must be given, and take advantage of, the time allocated to them to assess, understand, and utilize this knowledge in ATD decision-making.

The *selection framework* exists within a context of environmental and personal factors. For our purposes, we have defined *personal factors* to consist of resources (psychological, financial, and so on) that consumers (and their families and caregivers) bring to the process, as well as the skills and beliefs held by the relevant service providers. Thus, there are relevant environmental and personal factors associated with the consumer as well as environmental and personal factors associated with the provider that can influence the ATD selection process and outcome. Together, these environmental and personal factors create the context in which ATD decision-making and device selection for a given individual occur.

ENVIRONMENTAL PREDISPOSING ELEMENTS

Policies and priorities. Regardless of the country in which we are located, we live and work in a broad social environment consisting of expectations, priorities, and regulations that have resulted in large part from the enactment of laws and policy decisions at various levels of government. Laws and policies can affect the availability of ATDs (e.g., government standards for the approval of new products) and the ways in which they are made known and provided to consumers (AT service delivery). Applicable financial policies and resources for products, professional training, and service provision are frequently aligned with legislative mandates. Insufficient funding for ATDs, training for providers, the conduct of a comprehensive assessment, and training users and caregivers on obtained ATDs can result in inadequate resources to achieve a good match of person and technology (World Health Organization, 2011).

The role of culture. Culture is an important consideration in AT service delivery and is certainly and should be a component in problem-solving toward an AT solution Cook and Polgar (2012). Culture refers to specific patterns of behaviors and values (such as the provision of care) that are shared among members of a designated group and are distinguishable from those of other groups. Culture includes, but is not limited to, geographic origin, language, traditions, values, religion, food preferences, communication, education, and lifestyle. Individuals with a shared disability can have cultural identities, such as we see in Deaf Culture. The culture has a strong influence on our views of disability in general and of individuals with specific disabilities. It determines the value we place on the levels of independence and integration into society achieved by individuals with disabilities.

Cultural influences affect an individual's socialization, degree of exposure to technology, past experiences and satisfaction with technology use, and the consequent development of a favorable predisposition to the use of an ATD. For this reason, plus others, it is important when conducting an AT assessment to evaluate individuals on their perspectives regarding the use of technologies.

The manner in which AT resource allocations have been made has contributed to the underutilization of technology by some groups. If persons in low socioeconomic environments do not have technology, and are not accustomed to having it available, then this can become an expectation or assumption of a lifestyle that perpetuates that status. This may not be well understood by providers who believe that ATDs should be desired and used. Contrasts between belief systems may be observed during ATD discussions and decision-making. While providers may advocate for increased independence, some families may expect or be accustomed to increased dependence in the presence of a disability. The lack of a shared understanding may result in frustration and ultimately inappropriate ATD selection (De Jong, Scherer, & Rodger, 2007).

Expectations and attitudes of others. Cultural and subcultural values placed on functional support from assistive technologies, as opposed to personal assistance from family members, and the value families place on enhanced independence can affect an individual's predisposition to use an ATD. For example, there may be family resistance to ATDs that call attention to the user, which is perceived as "spreading stigma" to the family. Families may agree to the use of an ATD in the home and community but fail to follow through on this agreement when stigmatizing effects are perceived by the family (Scherer, 2005). Such attitudes are shaped by experiences with society in general, and they affect the willingness of individuals from culturally diverse backgrounds to work with ATD providers (World Health Organization, 2011).

Additionally, culture affects the manner in which family members perceive and respond to providers, and it exerts a strong influence on the way in which providers behave toward family members. Inherent in ATD decision-making is the assumption that partnerships among providers, family members, and the user are at least desirable, if not essential, in effectively identifying appropriate ATD solutions for use by persons with disabilities. These partnerships are more effective when consideration is given to cultural issues and to the characteristics and requirements of the milieu/environments where an ATD will be used (World Health Organization, 2011).

PERSONAL PREDISPOSING ELEMENTS

Resources. The aforementioned characteristics of the environment (or psychosocial milieu) have an impact on consumer's and provider's personal factors as they begin the process of ATD selection. Resources, both social and personal, may affect which ATDs will be recommended and obtained and also the extent to which an individual participates in community and societal roles. Thus, resources are strongly related to the overall success of rehabilitation efforts.

For consumers, important resources include emotional, physical, and material support from family as well as friends and other significant individuals. Support from others can promote such personal resources as enthusiasm, optimism, self-confidence, a sense of empowerment, and a readiness to try new approaches and products such as ATDs. In some countries, an individual's financial reserves become a key personal resource. In the United States, for example, a good many obtained ATDs are paid for out of pocket (World Health Organization, 2011).

The possession of resources in many cases is equated with options, informed choice, and a sense of empowerment and competence. Thus, resources are key to what the consumer is able to bring to the table during the selection of support (regardless if provided by an ATD, personal assistant, family help, or a combination of these) and type, style, and model of ATD.

Just as consumers differ considerably in their environmental and personal/psychological resources, so do providers vary in the resources they bring to the service delivery process. Some are equipped with an advanced degree and certification as an ATD provider. Some have a vast pool of contacts they can approach for help with complicated situations and questions or when confirmation is desired for a plan of action. Other providers may be more professionally isolated. Key resources for providers are those made available to them by their employers, the broader community, and what they obtained through their own creativity and initiative. Providers who possess excellent resources also enjoy a sense of empowerment, competency, and confidence.

Knowledge and information. Knowledge and information resident in the consumer, family member or caregiver, and provider, or having ready means to obtain information, is a strong influence on appropriate ATD selection (De Jong et al., 2007). While this can be considered a subset of the resources category discussed above, its importance is so strong as to warrant it being separately highlighted. For consumers and their significant others, access to computers and telephones can be key to being "connected" and in the information loop. In the

United States, support provided by independent living centers and through the state projects funded by the Assistive Technology Act has made a crucial difference in many consumers and caregivers having the knowledge and information they need to make informed choices. Providers depend heavily on conferences, professional literature and journals, and venders to supply them with current and detailed, technical, information about available products.

Expectations. Consumers bring expectations to the process that are internal and external. They carry the expectations of their culture, parents, spouses, children, employers, peers, and society in general. A society that expects an individual with a disability to strive to be as functionally and financially independent as possible conveys that message to the consumer either directly or indirectly. When employed, expectations are placed on that person for a particular level of job performance and independence in performing tasks and meeting objectives. Providers, too, have expectations placed on them by their employers. Third-party payers may demand documentation and paperwork typically above that mandated by any single employer.

Preferences and priorities. Beyond the expectations placed on consumers from external sources are expectations they place on themselves due to prior history with ATDs, their particular level of motivation, judgment, and outlook, and many other factors that serve to combine in a way defining each of us as unique individuals. These influences combined with personality and temperament characteristics (such as degree of self-determination and self-confidence, anxiety, and depressed mood) serve to determine our preferences and priorities, all of which have been linked to varied levels of ATD use and degrees of satisfaction (Cook & Polgar, 2012). It is important to note that these influences interact (or affect one another), and they can change with the passing of time and the accumulation of experience. Thus, at a given point in time, each consumer has a predisposition to view ATD use as being favorable or not, for certain purposes and in particular settings or environments.

When we think of the ultimate outcome of the ATD service delivery process, we think of a consumer satisfied with the use of a recommended device and who, in International Classification of Functioning, Disability and Health (ICF) terminology, is performing needed and desired activities and is able to participate in a variety of roles and events in varied settings where the lack of an appropriate device was a critical limiting factor for performance and participation. The perceived effectiveness of the ATD and the user's enhanced subjective well-being as a result of use are additional key outcomes. Unfortunately, such outcomes are not always achieved. It remains the case that too many consumers discontinue use of recommended ATD and that the reasons for this parallel those found in research studies of adherence and nonadherence to recommended healthcare interventions. Studies of a variety of interventions have demonstrated that the extent and nature of adherence problems are similar across diseases, across regimens, and across age groups and that a significant association exists between good adherence and personality factors, social issues, and consumer/patient beliefs (World Health

Organization, 2011). These factors, in turn, influence subjective well-being/quality of life, adjustment to the disability, concern about looking different and experiencing discrimination, readiness to work toward change, and willingness to adopt or comply with additional interventions, such as ATDs (World Health Organization, 2011).

It is important to assess the consumer's readiness for a given ATD and to provide the most appropriate device and training tailored to the unique characteristics and issues of that consumer. The ATD provider must be given, and take advantage of, the time allocated to them to assess, understand, and utilize this knowledge in ATD decision-making.

The consumer's arsenal of resources and knowledge, personal preferences and priorities, as well as expectations of the ATD selection process serve to define that person's predisposition to the use of a particular ATD. There is, therefore, a simultaneous *objective* or functional need for the ATD (consumer cannot walk 50 ft on a smooth surface) as well as the consumer's *subjective* predisposition to the use of a particular ATD (strong desire to independently move 50 ft on a smooth surface).

Predispositions to ATD use can be determined by assessing consumer preferences, knowledge, experiences, resources, and such characteristics as subjective quality of life/well-being, self-esteem, mood, and competence. These are the areas assessed by the Matching Person and Technology portfolio of assessments. Predisposition may be subsequently modified by a lengthy wait for ATD procurement and by ATD training and trial use (or the lack therefore), as well as during ATD selection which is the immediate focus. Furthermore, the *subjective need* for an ATD (i.e., consumer perceived need) often does not match *objective need* for an ATD as determined by providers and measures of functional limitation. The pressures placed on rehabilitative processes and the wait time inherent in ATD provision complicate meeting needs (Scherer, 2002).

The concomitant subjective views held by the ATD provider are no less important than are the candidate users' subjective perspectives on the selection of the most appropriate ATD. While consumers' predispositions are shaped by their resources, knowledge and information, expectations and preferences, and priorities, so too are the predisposition of providers. Therefore, it is important to assess the degree to which consumer and provider perspectives are shared, in addition to evaluating both the subjective and objective needs for an ATD. Otherwise, the path to dissatisfaction and nonuse of an ATD can begin at the selection stage. In other words, the predisposing elements for ATD selection will have an effect on the outcomes of ATD use (De Jong et al., 2007).

DEVICE SELECTION

People with disabilities around the world live, learn, play, and work in widely varying situations or communities. The United Nations Convention on the Rights of Persons with Disabilities and its Optional Protocol need to be operationalized within a context of economies of varying strength and infrastructure. One important means by which individuals with disabilities can function more independently is with the support of ATDs (also known as assistive products—ISO 9999), which can make it possible to pursue education, employment, and involvement in community life. Yet, the barriers and individuals' views of the value of ATDs in accomplishing their objectives, and their predisposition to use one or more ATDs, vary considerably. Such variation is due to a variety of personal and environmental/social factors reflecting the resources and perspectives brought to the ATD decision-making process by both consumers and providers. These factors can be identified and measured, and then used to guide device selection in situations where there is the potential for a poor match of user and technology and for an unsuccessful outcome.

SUMMARY

Although a wide range of ATSD systems exists around the world, they have one thing in common: to ensure that people receive the ATDs and services they need when they need them. To achieve safe, effective, and sustained use, it is important that the selected ATD matches the user and the environments of use, that the user receives adequate training and followed up services, as well as the necessary servicing and maintenance of the ATDs provided. Environmental and personal factors associated with the user, the provider, and the state or country can influence the selection and outcomes associated with ATDs.

REFERENCES

Cook, A. M., & Polgar, J. M. (2012). *Essentials of assistive technologies.* St. Louis, MO: Mosby.

De Jong, D., Scherer, M., & Rodger, S. (2007). *Assistive technology in the workplace.* St. Louis, MO: Mosby.

Eldar, R., Kullmann, L., Marincek, C., Sekelj-Kauzlaric, K., Svestkova, O., & Palat, M. (2008). Rehabilitation medicine in countries of Central/Eastern Europe. *Disability Rehabilitation, 30,* 134–141.

Field, M. J., & Jette, A. M. (2007). *Committee on disability in America: The future of disability in America.* Washington, DC: National Academies Press.

Horizontal European Activities in Rehabilitation Technology Consortium. *HEART final report on service delivery.* (1995). Retrieved from ⟨http://portale.siva.it/files/doc/library/a416_1_ATServiceDelivery_HEART_ReportC51.pdf⟩.

Øderud, T., Brodtkorb, S., & Hotchkiss, R. (2004). *Feasibility study on production and provision of wheelchairs and tricycles in Uganda.* SINTEF Health Research. Retrieved from ⟨http://siteresources.worldbank.org/DISABILITY/Resources/Regions/Africa/311048-1239047417106/WheelchairsUganda.pdf⟩.

Oderud, T., & Grann, O. (1999, November). Providing assistive devices and rehabilitation services in developing countries. In: *The 5th European conference for the advancement of assistive technology*. Düsseldorf.

Scherer, M. (2002). Introduction. In M. Scherer (Ed.), *Assistive technology: Matching device and consumer for successful rehabilitation* (pp. 3–13). Washington, DC: American Psychological Association.

Scherer, M. (2005). *Living in the state of stuck: How assistive technology impacts the lives of people with disabilities* (4th ed.). Brookline, MA: Brookline Books.

Shay, A. F., Anderson, C. A., & Matthews, P. (2017). Empowering youth self-definition and identity through assistive technology assessment. *Vocational Evaluation and Work Adjustment Association Journal*, *41*(2), 78–88.

Tinney, M. J., Chiodo, A., Haig, A., & Wiredu, E. (2007). Medical rehabilitation in Ghana. *Disability Rehabilitation*, *29*, 921–927.

United Nations. *Production and distribution of assistive devices for people with disabilities*. (1997). Retrieved from ⟨http://www.dinf.ne.jp/doc/english/intl/z15/z15001p1/z1500101.html⟩.

World Health Organization. (2005). *Fifty-Eighth World Health Assembly. Provisional agenda item13.13: Disability, including prevention, management and rehabilitation*. Geneva: Author.

World Health Organization. (2011). *World report on disability*. Geneva: Author.

Referral, intake, and assessment

7

Marcia Scherer

Institute for Matching Person and Technology, University of Rochester Medical Center; Physical Medicine and Rehabilitation and Senior Research Associate, International Center for Hearing and Speech Research, Webster, NY, United States

In the previous chapter, *Overview of the Assistive Technology Service Delivery Process: An International Perspective*, the process was explained as a system with many interacting components that vary over time and by place. Consider the chart below. It depicts five aspects representing priority research themes discussed during the World Health Organization's August 2017 Global Research, Innovation, and Education in Assistive Technology (GREAT) Summit at its headquarters in Geneva (World Health Organization, 2017). Coordinated by WHO's Global Cooperation on Assistive Technology (GATE), the GATE priority research themes form the core around which the Summit discussions focused. (A forthcoming special section of the journal *Disability & Rehabilitation: Assistive Technology* will feature five original articles on each of the priority research themes as well as one summarizing content presented during an electronic poster session. Updated information can be found on the journal's website: https://www.tandfonline.com/loi/iidt20 as well as on the website for the World Health Organization's GATE: http://www.who.int/phi/implementation/assistive_technology/en/.)

As stated in the previous chapter:

> *Assistive technology (AT) service delivery takes place within an AT system. The components of this system include users and their families, AT products, AT services, personnel, service providing agencies, manufacturers, funding agencies, and policies and legislation. The components of the AT systems vary substantially within and across countries, as do the interactions between the components. For example, some manufacturers work closely with users on the design of assistive products while others do not; in some countries a broad range of products and services are available while in others they are limited; and in some contexts the national or local government carry the cost for AT, in other contexts users and their family members need to pay fully for them, and in yet other contexts they are paid for through donations (p. 2 of Word version).*

Assistive Technology Service Delivery. DOI: https://doi.org/10.1016/B978-0-12-812979-1.00007-2

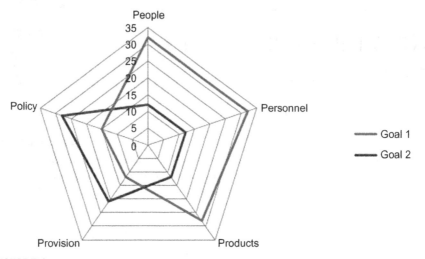

FIGURE 7.1

Five GATE priority research themes: example of establishing a new AT service model. *GATE*, Global Cooperation on Assistive Technology.

Referring to Fig. 7.1, the five GATE research priority areas of people, personnel, provision, products, and policy are interconnected but vary in strength according to the goal established. For example, Goal 1 may be to increase the number of people provided with appropriate AT in a given vocational rehabilitation office. Adequate policies are in place, a model of provision exists, but there is a shortage of trained personnel and dearth of knowledge about affordable products. People served, personnel availability, and product knowledge thus are revealed as areas of highest priority in achieving Goal 1. Now let's consider that Goal 2 might be the establishment of an integrated interdisciplinary network or team by the agency and those with whom it has associations and collaborations. Personnel and their knowledge of products exist, people exist who need to be served and provided with the most appropriate AT, but new policies for such an integration and means of provision need to be agreed upon and established, making the last two themes the highest priority of the five.

Continuing with the example of Goal 2 and the establishment of an integrated interdisciplinary network or team, a new system for referral and intake needs to be developed or modified. What follows is a discussion of examples of key considerations important to take into account.

REFERRAL

Our hypothetical new service will receive external referrals, make external referrals, and make internal referrals to particular specialists on the team. In the general medical literature, there are numerous articles describing the breakdowns and inefficiencies that have been found to occur in all components of the specialty-referral process (e.g., Ghandi et al., 2000; Mehrotra, Forrest, & Lin, 2011). Surveys of primary care and specialty physicians indicated that major issues were lack of timeliness and inadequate content and information. On average, specialists reported that they did not receive enough information to adequately address the problem 23% of the time (Mehrotra et al., 2011). Key issues needing more details were problems to address and questions to answer. Primary care physicians wanted answers to specific questions from specialists, the specialist's assessment of patient including results of test and procedures, and the recommended treatment course (therapy proposed or initiated).

Physicians and medical referrals are likely not alone. Referrals between and within all allied health professions, vocational rehabilitation, etc. would no doubt receive similar reviews. Although electronic recordkeeping may have a positive impact on this situation, not enough is known at this point in time (e.g., Kim, Chen, & Keith, 2009) to state this with any degree of certainty.

There are fundamental bits of information needed for an effective and referral regardless of profession: primary need, desired outcome, and urgency come immediately to mind. Other key information can be obtained from an intake, and to that topic we now turn.

INTAKE

The late George Engel, former physician at the University of Rochester Medical Center and generally acknowledged as the founder of the biopsychosocial model, believed that we need to listen more to the patient and the patient's history because therein lies the diagnosis. In an article on creating a "bulletproof letter of medical necessity," Davin (2015) stated,

> The importance of developing a complete medical history is key. The evaluating therapist should not only acquire a list of diagnoses within the client's medical history, but also carefully assess other factors that parallel a diagnosis, such as pain, upper-extremity weakness, poor trunk control, or generally decreased function, for it may be these other factors, often referred to as secondary diagnoses, that play a larger role in determining the equipment that will be selected.

Davin advises emphasizing the impact of the identified limitations on the person's everyday activities as well as the inclusion of such supporting documentation as "physician's notes and supplementary data (i.e., photographs, standardized testing, etc.)." Also helpful to note are observations about supports available, accessibility of the settings of use (including agreement of significant others), the person's documented desire to and readiness for use of the AT, and possession of the necessary skills for use.

To obtain this information, a systematic assessment process is needed beginning at intake with the definition of the relevant issues and priorities. The next section presents one such process.

ASSESSMENT

There are now thousands of assistive, access, educational, and workplace technologies and a lot of choice available within any given product line. Achieving an optimal match of person and technology can be a complex process. How do we start to narrow that down? An entire book could be written on AT assessment and, in fact, one has (Federici & Scherer, 2018). Additionally, in a series of books referred to as the "Matching Person & Technology (MPT) trilogy," plus others, a consistent process is presented. The relevant books are as follows:

Scherer (2005). *Living in the State of Stuck: How Assistive Technology Impacts the Lives of People with Disabilities, Fourth Edition*. Cambridge, MA: Brookline Books. (Focuses on adults with cerebral palsy and spinal cord injury and their use or nonuse of technologies for activities of daily living, mobility, and communication.)

Scherer (2004). *Connecting to Learn: Educational and Assistive Technology for People with Disabilities*. Washington, DC: American Psychological Association (APA) Books. (Focuses on children and adults with hearing and visual impairments.)

Scherer (2012). *Assistive Technologies and Other Supports for People with Brain Impairment*. New York: Springer Publishing Co. (Focuses on products for those with cognitive disabilities through the lifespan: born with a cognitive disability, have adult onset stroke or traumatic brain injury (TBI), and who developed dementia.)

EMPLOYMENT

de Jonge, Scherer, and Rodger (2007). *Assistive Technology in the Workplace*. St. Louis, MO: Mosby.

Material from these books has been extracted and used to develop the following succinct summary.

The Matching Person and Technology model

Fig. 7.2 depicts a bull's eye chart. While it addresses AT, it is equally applicable to access, educational, and workplace technologies and it depicts the complexity of the process of achieving a good match of person and technology—our target or bull's eye. A logical place to begin is by addressing the person's unique characteristics and resources of that individual person. What are the person's needs and goals? What are the typical routines of the person at work and home and what

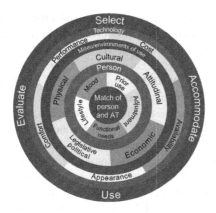

FIGURE 7.2

The MPT bull's eye chart. *MPT*, Matching Person and Technology.

presents difficulties for them? What interferes with the person achieving what they want, going where they want to go, and doing what they want to do? Is the person generally happy and composed? Or sad and anxious? What is the nature of the person's need or difficulty? What is the person using or doing now? What does the person believe they want and need to do in order to succeed in employment? Does the person prefer to have someone else help them because they prefer that interpersonal contact? What is that person's receptivity or predisposition to the use of a technology?

Moving outward from the center of the circle and beyond the characteristics and resources of the person, considerations related to the characteristics and requirements of the environment(s)/milieu of use and their impact on the individual learner become key.

The word *milieu* is used because it connotes that fact that our environment is not just a built one consisting of physical objects and electrical outlets and lights, but a place comprised of people who have a variety of attitudes and values. What are the predispositions of others to using a technology with this person? Will the family encourage and support use? Often caregivers are as much primary users of these technologies as the persons. And it's important to assess their perspectives as well as those of the persons. Are all of the necessary supports in place for this person to use the planned technology? What can we do in the physical or built environment to help this person function better and then what can we do to help that person develop a sense of belonging and connection?

A factor that needs to be considered when evaluating technology preferences is the increasing diversity of cultures, languages, and lifestyles in the United States. Changes in the traditional family structure, in the cultural backgrounds, and in the linguistic heritages of families necessitate the need to identify resources and develop skills in working with families from diverse backgrounds and understanding their perspectives regarding the selection and use of technology.

Characteristics of the technology being considered is the next circle as we move outward from the bull's eye. How available is it? How much does it cost and who will pay for it? Are there effective alternatives that cost less? What's the performance and reliability record of the technology? Repair records? Maintenance requirements? Will the vendor come in and provide services for it? Is the technology comfortable and easy for the person to use? How does the person feel using it? Is it important that it be upgradeable and how easy are upgrades to get? What is the impact of climate on this technology? Humidity, heat, cold? If it has to be portable, is it? How durable is it and can it withstand a lot of wear and tear in going from room to room or home to work? How compatible is it with other equipment being used on the job and at home? Is the person already using a device or number of devices, and will it interface well? Or is a point of overload being reached? How accurate and fast is the device? Is training needed in order for the person to use this device?

The outermost ring is when we start to narrow down and prioritize choices and options. With a good, general idea of the desired technology, we want the person to see the device and use it on a trial basis so that we can finalize the most desired features. Should the technology be leased, purchased, or rented? Will it require customizing or other adaptations to the person?

One of the person uses it in actual situations and natural settings, and feedback is obtained on how well the technology is performing for that person, then consideration might be given to selecting a peripheral or additional or ancillary devices, making the necessary adjustments and upgrades and so on until the person is preparing to graduate and a transition plan is developed with vocational needs in mind.

Through this process, prioritize the options. Document why it is that one product or feature is preferable to another. Document that you've taken into account the person's changing needs and that the device will be used as intended. And when you do your follow-up, document that it is in fact being used as intended. If the person is supposed to use it 50% of the day, but only uses it 25% of the day, then that needs to be addressed. Why is that happening? Does the person have changing needs that need to be evaluated?

Undeniably, this is a lot of information to assess and obtain. An assessment process consisting of a series of instruments has been developed to guide just such information-gathering and assist in the organization of the many influences which impact on the use of assistive, access, and educational technologies—including psychological and social factors. This information is then balanced with the characteristics of the environment in which the technology will be used along with the features and functions of the technology itself. The Matching Person and Technology (MPT) process is both a practical and a research tool which identifies barriers to technology use for a particular individual and its use has resulted in high satisfaction with options that match not only the individual's strengths and needs but also preferences and temperament.

Use of a systematic, evidence-based process for decision-making, AT selection, and follow-up

This chapter began with Figure 1 as a depiction of five priority research themes discussed during the World Health Organization's August 2017 GREAT Summit (World Health Organization, 2017). Two goals served as examples:

Goal 1: Increase the number of people provided with appropriate AT in a given vocational rehabilitation office.
Goal 2: Establish an integrated interdisciplinary network or team.

A process was described in the previous chapter in the section, *The ATD Decision-Making and Device Selection Process* (starting on page 9 of the Chapter 6 Word version), as a means of achieving these goals. In Figs. 7.3 and 7.4, the process and its components are depicted in graphic form and text to show how they dynamically evolve. Additionally, Fig. 7.5 illustrates the assessment forms used in the process.

Form 1. Initial Worksheet for the Matching Person & Technology Process

This form is organized by areas in which persons may experience loss of function (e.g., communication, mobility, hearing, and vision). It is designed to be used by providers working together with persons to identify areas to strengthen through

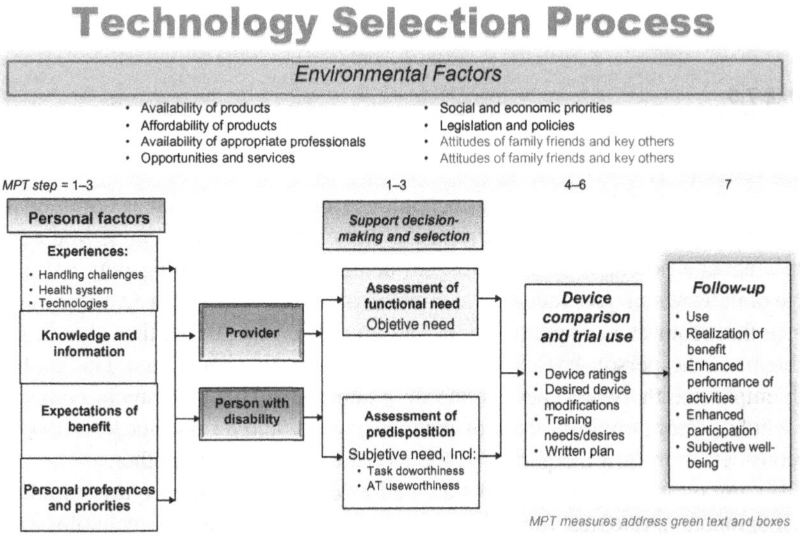

FIGURE 7.3

The technology selection process.

Step 1	Initial worksheet
Step 2	History of support use
Step 3	Specific technology matching (assistive, educational, workplace technologies, and more)
Step 4	Identification of factors that may indicate problems with acceptance of the technology or realization of benefit from use
Step 5	Identification of intervention strategies
Step 6	Plan and documentation
Step 7	Follow-up

FIGURE 7.4

Process and portfolio of assessments.

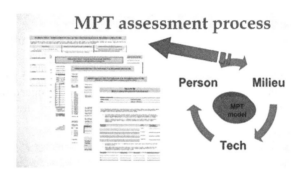

FIGURE 7.5

Assessment forms used in the process.

the use of an assistive (or other technology) or environmental accommodation. While this forces a focus on the "separate parts" of the person, unless you address each area as a potential obstacle to goal achievement, something may be missed. For example, when you focus on communication and are about to recommend a device that requires very good vision, and you haven't assessed that, there may be problems if the person has experienced some form of vision loss. The goal is to both emphasize the whole person and do a comprehensive assessment considering the whole person, environments of technology use, and so on, but to achieve this by considering in turn the parts and their relationship to one another.

It is important to note strengths as well as difficulties. Identifying initial goals and beginning strategies for achieving them may focus on a technology, or a change in the environment, or both. Each item should be addressed, regardless if you think it's relevant for this person or not.

Form 2. History of Support Use

The History of Support Use form is used to understand technologies that have been attempted, their success, and why a new technology may be better. The organization of this form is consistent with areas of functioning found in the Initial Worksheet for the MPT process. It includes space for listing three technologies (or supports/strategies) that have been tried for each area.

Forms 1 and 2 focus on a person's separate areas of functioning, given the belief that key obstacles to optimal technology use are identified only when each area is considered. For example, if a focus on communication leads to a recommendation for a device that requires good vision, problems may be encountered when using the recommended device if vision has not yet been assessed. A goal that emphasizes the need to focus on the whole person is achieved by considering the many parts that comprise the whole and their relationship to one another.

Form 3. Survey of Technology Use (SOTU)

After a technology is determined to be viable, the individual is asked to complete the Survey of Technology Use (SOTU), a 29-item checklist that inquiries into the respondent's present experiences and feelings toward technologies. This information is gathered and evaluated in an attempt to identify and introduce new technology that builds and capitalizes on existing skills and comfort in their use. Respondents also provide information about their general mood, personal characteristics and preferences, and social involvement. These areas have been identified through research to impact a favorable predisposition toward technology use.

The SOTU provides two identical forms, one for use each by professionals and one by consumers. Both forms are meant to be used as a set, per Figure 3, providing different perspectives from the professional and consumer that may need to be addressed.

Form 4. Assistive Technology Device Predisposition Assessment (ATD PA)

The Assistive Technology Device Predisposition Assessment (ATD PA) (Person and Device Forms) has supporting materials (Figs. 7.6 and 7.7): (1) computerized scoring and interpretations and (2) an interactive CD training program for professionals. The Person Form has 54 items divided into three sections. Section A (9 items) asks for consumer ratings of functional capabilities, Section B (12 items) provides information on subjective well-being in the context of the World Health Organization's International Classification of Functioning, Disability and Health (ICF) domains of Activity and Participation, and Section C (33 items) produces data on personal and psychosocial factors through eight subscales that assess mood, self-esteem, self-determination, autonomy, family support, friend support, therapist and program reliance, and motivation to use support.

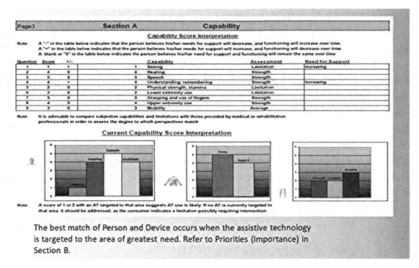

FIGURE 7.6

Sections of the Assistive Technology Device Predisposition Assessment (capability score).

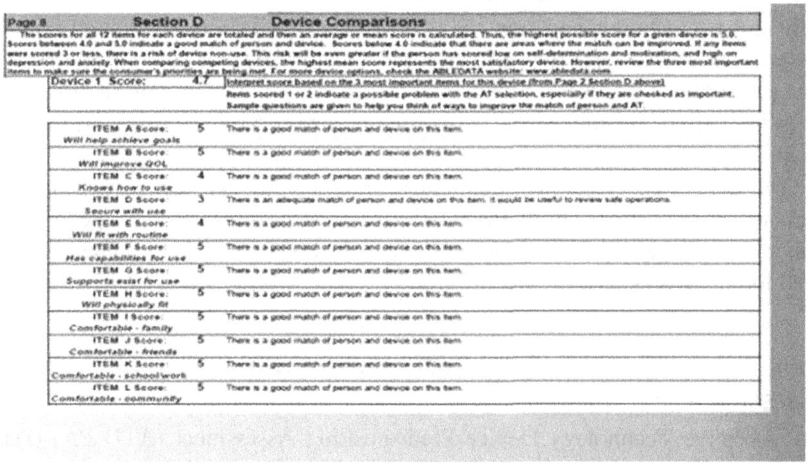

FIGURE 7.7

Sections of the Assistive Technology Device Predisposition Assessment (device comparison).

The Device Form's 12 items ask respondents to rate their predisposition to using the specific AT under consideration. A follow-up version of this form exists to assess satisfaction with use of the selected device.

The professional forms of the ATD PA allow them to determine and evaluate incentives and disincentives to the use of the device by a particular consumer and to compare perspectives held by the consumer and the professional.

Form 5. Educational Technology Device Predisposition Assessment (ET PA)

The Educational Technology Device Predisposition Assessment is a 43-item self-report checklist developed to assist professionals and teachers in compiling comprehensive pre- and postlearning profiles of a student provided with educational technology to determine whether its use enhances the student's educational experience. These profiles can be used to help demonstrate improvement in skills for individual students and organize information about the needs of a particular student population.

A companion teacher/educator version of this form allows for an assessment of the view of both a student and his or her teacher in four key areas: characteristics of the educational goal and need that a teacher is attempting to address through the use of a specific technology, characteristics of the particular educational technology being reviewed, characteristics of the psychosocial environments in which the technology will be used (e.g., the presence of supportive family, peers, and/or teachers), and the student's characteristics that may influence technology use (e.g., learning styles and preferences).

Other measures in the MPT set include the *Workplace Technology Device Predisposition Assessment, the Healthcare Technology Device Predisposition Assessment, Hearing Technology Predisposition Assessment, Cognitive Support Technology Predisposition Assessment, Matching Assistive Technology & Child-ACES* for Early Intervention and ages 0–8, and *Matching Student and Technology* for children and adolescents in inclusive education. There are follow-up versions of each of the measures.

Several MPT measures are suitable for use with the World Health Organization's ICF and thus are relevant for use in assessing ICF domains impacted by technology use. It is compatible with the biopsychosocial model and helps implement Dr. George Engel's philosophy of listening to the patient and that person's history, experiences, and priorities. Because of the many components addressed, it also can well serve the need to create better referral documentation as it helps paint a full picture of the person's current physical and psychosocial situation and desired goals. As stated by Davin, "Remember, the more detailed picture you paint, the better understanding the funding source reviewer will have of the client, and it perhaps will better facilitate the approval process" (Davin, 2015). The results of the MPT process will guide the service provider and team into addressing the following seven questions:

1. Which aspects of the person's abilities and capabilities serve as strengths and which pose obstacles to what the person wants to do and accomplish?
2. What does the person want to do that cannot be done as they wish?
3. What personal characteristics and factors are strengths for goal achievement and which serve as hindrances?
4. Once a support is selected, are expectations of benefit positive? Realistic?
5. What raining may the person require to achieve maximum benefit from the support? Is that training available? Affordable?

6. What additional supports would be useful?

7. Upon follow-up, has the person realized benefit from use of the support? To what extent? Is an upgrade or additional support needed?

Above all, the outcome of the process will result in the identification of most appropriate AT for the unique individual and document the variety of reasons why this is the case.

The MPT model is a validated resource translated into multiple languages, is a major component of the key service delivery models (Bauer, Elsaesser, Scherer, Sax, & Arthanat, 2014), and is used in service delivery research projects (e.g., Craddock & McCormack, 2002; Hendricks et al, 2015; Minton et al, 2017). It is the model organizing the Assistive Technology Assessment Handbook by Federici and Scherer (2018).

SUMMARY

This chapter has presented a dynamic information-gathering process where the provider will learn about a person's unique history, perspectives, needs, and priorities. The process provides problem-solving ways to address challenges and barriers that are identified. What we want to avoid in AT service delivery is to have a piece of AT in mind and then work backwards to show the person why this is the appropriate device to use. There's nothing wrong with looking ahead and forming hypotheses that certain pieces of AT may be appropriate; but it is important to ensure that any consideration of technology goes directly back to what it is individuals are trying and needing to do, their interests and personal characteristics, and the particular characteristics of the AT selected. There are great many AT products available, and individuals can exercise considerable choice, but the outcome will depend upon how well the device has been matched to the unique person, and not vice versa.

REFERENCES

Bauer, S., Elsaesser, L. J., Scherer, M., Sax, C., & Arthanat, S. (2014). Promoting a standard for assistive technology service delivery. *Technology and Disability*, *26*(1), 39−48.

Craddock, G., & McCormack, L. (2002). Delivering an AT service: A client-focused, social and participatory service delivery model in assistive technology in Ireland. *Disability and Rehabilitation*, *24*(1−3), 160−170.

Davin, K.N. (July 29, 2015). Creating a bulletproof letter of medical necessity. Rehab Management. Retrieved February 20, 2018 from: ⟨http://www.rehabpub.com/2015/07/creating-bulletproof-letter-medical-necessity/?_hsenc = p2ANqtz−hMQ4_9S7P0hhk AgaXXHL4iX_xXd_SZtdKFAk5q5iDlFlZdsgvdSZ0mHZWGWNOBoq-45IeGLuA3 TzsDhzSsYDhcyqcA&_hsmi = 61321590⟩.

de Jonge, D., Scherer, M. J., & Rodger, S. (2007). *Assistive technology in the workplace.* St. Louis, MO: Mosby.

Federici, S., & Scherer, M. J. (Eds.), (2018). *Assistive technology assessment handbook* (2nd ed.). Boca Raton, FL: CRC Press.

Gandhi, T. K., Sittig, D. F., Franklin, M., Sussman, A. J., Fairchild, D. G., & Bates, D. W. (2000). Communication breakdown in the outpatient referral process. *Journal of General Internal Medicine, 15,* 626−631. Available from https://doi.org/10.1046/j.1525-1497.2000.91119.x.

Hendricks, D. J., Sampson, E., Rumrill, P., Leopold, A., Elias, E., Jacobs, K., ... Stauffer, C. (2015). Activities and interim outcomes of a multi-site development project to promote cognitive support technology use and employment success among postsecondary students with traumatic brain injuries. *NeuroRehabilitation, 37*(3), 449−458. PMID: 26484525.

Kim, Y., Chen, A. H., Keith, E., et al. (2009). Not perfect, but better: primary care providers' experiences with electronic referrals in a safety net health system. *Journal of General Internal Medicine, 24*(5), 614−619. Available from https://doi.org/10.1007/s11606-009-0955-3.

Mehrotra, A., Forrest, C. B., & Lin, C. Y. (2011). Dropping the baton: Specialty referrals in the United States. *Milbank Quarterly, 89*(1), 39−68. Available from https://doi.org/10.1111/j.1468-0009.2011.00619.x.

Minton, D., Elias, E., Rumrill, P., Hendricks, D. J., Jacobs, K., Leopold, A., ... Scherer, M. (2017). Project career: An individualized postsecondary approach to promoting independence, functioning, and employment success among students with traumatic brain injuries. *Work, 58/1,* 35−43. Available from https://doi.org/10.3233/WOR-172598.

Scherer, M. J. (2005). *Living in the state of stuck: How assistive technology impacts the lives of people with disabilities* (4th ed.). Cambridge, MA: Brookline Books.

Scherer, M. J. (2004). *Connecting to learn: Educational and assistive technology for people with disabilities.* Washington, DC: American Psychological Association (APA) Books.

Scherer, M. J. (2012). *Assistive technologies and other supports for people with brain impairment.* New York, NY: Springer Publishing Co.

World Health Organization. (2017). *Great summit report.* Geneva: WHO; Licence: CC BY-NC-SA 3.0 IGO.

Plan development, recommendations, and report writing

8

Anthony Shay

Capacity Building Specialist, Assistive Technologist, and Rehabilitation Specialist, University of Wisconsin-Stout Vocational Rehabilitation Institute (SVRI), Menomonie, WI, United States

FOLLOWING THE ASSESSMENT . . .

The assessment process is designed to gather information—functional skills, purpose for the referral, and the consumer's (and referral/funding source) goals. This helps the assistive technology (AT) professional to better understand the consumer's needs relative to the AT they will recommend. The objective is to mitigate or eliminate the functional limitations keeping them from effectively engaging in the specific work activities within the specific work context(s) in which they must be performed. The consumer's overarching *goal* is the employment goal they have selected with the disability-employment professional (for the *purpose* of finding gainful employment).

The referral/funding source for AT services works with a consumer to determine a vocational goal (*Note:* AT services may also be provided to allow a consumer to participate in the disability-employment process). A referral for AT services may be received prior to a consumer having identified an employment goal. However, in disability-employment services, the focus, as a matter of course, is to develop an employment goal. At some point in the AT professional's interactions with the consumer, the conversation will center on vocational interests. There are occasions when a disability-employment professional recognizes a consumer may have functional limitations which will impact, at the very least, their ability to engage in job seeking activities as well as the interview process and job maintenance following acceptance of a position. AT referrals are appropriate before and after employment goals are determined.

THE EMPLOYMENT GOAL

Contrary to popular belief, there is no vocational lodestar—no single indicator of the vocational goal we should develop into a job. Think of the Great 8 (discussed in Chapter 1: Accommodation System: Self of this text) as your vocational constellation. You look up into the night sky trying to discern the points of vocational light that will outline the job goal on which you should be focused. Each point of light you see in the vocational constellation taking shape before your eyes is represented by your Great 8: strengths, resources, priorities, concerns, abilities, capabilities, interests, and informed choice. Now imagine that each of these points of light vary in intensity depending on the distance the light must travel to reach your eyes. Focusing on only one or two points of light (maybe just the brightest stars) and ignoring the others can have major goal attainment ramifications. Imagine spending your time, effort, and resources on becoming a doctor or nurse only to realize after graduating and looking for work in the medical field that you cannot stand the sight of blood, or working to become a secondary school teacher only to realize later that you dislike being around children. We must consider as many aspects of an employment goal (the vocational "stars") as possible to avoid preparing us for a job we will find little meaning and satisfaction in once we become employed.

Internationally renowned psychologist Mihalyi Csikszentmihalyi makes the point that it is easy to disregard obvious facts choosing to strive toward goals we cannot attain which would not satisfy us even if we were able to reach them. He finds that "what we can do well, that we enjoy doing, and that there is a demand for" is a worthwhile endeavor. He further asserts that when we happen upon a worthwhile endeavor "one should try to explore as many of one's abilities as possible so as not to miss any hidden potentialities" (Csikszentmihalyi, 2003, p. 173). Haworth points out that well-being is derived from employment which provides the following five qualities:

- Time structure;
- Social contact;
- Collective effort or purpose;
- Social identity or status; and
- Regular activity.

The transactional nature of the employment experience imposes these qualities on workers as a matter of pragmatic functionality, where these qualities are lacking well-being deteriorates. Well-being manifests when some degree of each of these qualities are present (Haworth, 1997, p. 24). If we accept this premise, then our personal attributes, resources, and the inherent qualities of employment comingle and define the satisfaction we achieve while working.

When we begin planning for a new job or a new position with a current employer, we must work to avoid fixating on that single point of light. The allure of the one or two brightest lights in our vocational constellation has the effect of blinding us to the others in our night sky. We miss the forest for the trees. The consequence of this is finding ourselves in a job we would prefer not to have while brooding over the sacrifices it took for us to get there. We then find only extrinsic motivators or reasons to remain in the job (e.g., a paycheck, insurance) having failed to build in the elements that lead to satisfaction and meaning (i.e., internal motivators). We sound the job's death knell the moment we accept a position devoid of intrinsic motivators. The job's days are numbered, and the job search will begin anew. Although it is not unheard of for people to hold poorly fitting jobs for long periods of time, it takes a toll on them. They tend to become bitter, angry, and negative people. When considering an employment goal, it is important to consider whether workplace accommodations, such as AT, might facilitate overcoming the barriers to employment caused by our functional limitations. We want to be aware of the Great 8 without quashing a consumer's goals before considering whether they can perform the essential functions of the job they want to pursue with or without accommodations. Once a consumer has identified a potential job goal, they may engage in employment services to determine their goodness of fit with the type of work they have chosen. Determining the potential need for accommodations is an important component in the goal development process.

FRAMING PROGRESS MEASURES: SMART GOALS

SMART is an acronym which stands for specific, measurable, attainable, realistic, and timely. It is always a good idea to ensure AT recommendations (AT outcomes or deliverables) are structured for clarity and so they are measurable; consumers need to know exactly what expectations have been placed on them and so, progress can be monitored toward goal attainment. Through establishing progress measures and goals, we frame the reasons and purpose for engaging in activities. It provides a means by which we are able to measure progress against reaching the goals we set for ourselves. The SMART process helps us craft progress measures and goals to make the expectations inherent in them clear and concise, leaving little room to question whether we are on-track to meeting them or when we have attained them. SMART goals are

- Specific: In Chapter 17 of this text, we discussed being specific regarding case noting for clarity. When we establish goals, we need clarity as well. Consideration (a future-oriented variation) of the five W's and three H's works well here also:
  - **Who** is the person or are the people being referred to or are responsible for or taking on specific roles related to the case activity?
  - **What** exactly is the activity to be performed—what are the steps to be undertaken?
  - **When** is the specific timeframe for the activity? When will it occur?
  - **Where** will the specific activity occur? Include multiple contexts and transit between contexts.
  - **Why** is the activity taking place? Provide the rationale behind the need for the activity.
  - **How** will the specific activity take place?
  - **How** much will the specific activity cost? Is it reasonable, necessary, and appropriate for the activity?

- **How** long will the activity take? What is the total expected duration of the activity?
- Measurable: Measurable goals provide an answer of the question "How will I know if I have achieved my goal?" We create the metrics by which we track our progress and which serve as pacesetters for our activities considering deadlines and timeframes for action.
- Attainable: When a goal is attainable, it is achievable. As discussed in Part 1 of this text, we balance the skills we possess with the challenges we face while engaged in tasks. Maintaining a balance (trying to achieve a higher average each time we engage in the task) influences our motivation and satisfaction related to goal striving. We may set intermediate goals based on task feedback toward attainment of our overall goal.
- Realistic: Goal-striving behavior is purpose driven. Goals are relevant to the purpose for which we create them. Realistic goals are representative of our alignment, vision, and engagement in tasks relevant to the purpose for which we establish goals.
- Time-bound: Building a timeframe into the goals we set primes us for goal-directed behavior. It facilitates a sense of timeliness, importance, and urgency around a goal that enables achievement (University of Maine System, 2018).

Example of an unSMART goal: I will obtain AT for work.

Example of a SMART goal: I will participate in an AT assessment that will allow me to identify potential AT devices or strategies toward gaining employment in the accounting field by July 20.

ASSISTIVE TECHNOLOGY PLAN, RECOMMENDATIONS, AND REPORT

The AT plan is typically written immediately following the AT assessment (on-site with the consumer immediately following the assessment) while the assessment, trials, demonstrations, simulations, and discussions regarding device characteristics and features are still fresh on the consumer's mind. Homework may be assigned as a follow-up (maybe assigned at the intake meeting to prime consumer interest, curiosity, and spur innovation). However, plans may be written following the assessment and sent to the consumer to review, sign, date, and return to the AT professional. Signatures are collected as an acknowledgment of agreement with the plan. An AT plan may include the goal(s) of the assessment, the employment goal of the consumer, the context(s) in which the AT is expected to be used, a listing of the consumer's functional skills and limitations (i.e., the functional skill/limitation area(s) being addressed), the methods used toward determining AT characteristics/feature matching, and the outcomes or recommendations to be included in the report. Items included in the AT Plan include

- Contact information for the consumer;
- Contact information for the referral/funding agent and agency;
- Contact information for the AT service professional and agency;
- Consumer's employment goal;
    - Short-term employment goal (e.g., interim job/work study while in school)?
    - Long-term employment goal?
- Consumer background information;
    - Context(s) for AT use (existing or potential accommodations);
    - Expected work tasks;
- SMART Progress measures;
- Expected timeframes (including consumer priorities);
- A list of the referral/assessment questions addressed in the AT assessment (the AT focus areas addressed in the assessment). During the assessment, additional needs may have been uncovered and included as functional areas to be addressed. This should be made clear in the assessment and recommendations to the referral/funding source.
- Device demonstration(s) or simulations performed;
- Whether AT was trialed;
- If a device loan program was utilized; and
- Outcomes (specific recommendations).

The AT recommendations and report are driven by the AT assessment process. This facilitates a determination of consumer needs and endeavors to match these needs to assistive technologies which facilitate task engagement in the identified context in which the AT will be used. The AT professional considers input from the consumer and other members of the rehabilitation team who may have been present, structures, and organizes AT recommendations for the consumer into a document outlining the agreed-upon outcomes. Recommendations should be viewed both individually and as a whole. They may be stand-alone products or part of an AT system. In any event, recommended AT may seem out of place, inappropriate, or insignificant by itself (whether part of an AT system or not) until viewed in conjunction with other recommendations. The report summarizes the assessment and formalizes the recommendations structuring and organizing them around the referral/assessment question, assessment goals, methodology for determining device characteristics/features, and the associated outcomes or recommendations. In addition to the bulleted points above from the AT planning process, the report may include the following elements:

- AT assessment summary;
- Consumer background information;
    - History of accommodation use;
    - Consumer disability/functional limitation history and current considerations;
    - Involvement of others (coworker/supervision issues, job coach, personal cares, task shifting, or sharing);

- Discussion of any AT items the consumer is currently using or has used in the past and their impact on meeting consumer needs for the purpose of the referral;
- Accommodation issues and strategies discussed;
- Specific AT devices or systems discussed;
  - Pros and cons;
  - Costs;
  - Repair and warranty issues;
- A list of the functional skills/limitations being addressed. Generally organized by
  - Sensory skills;
  - Physical skills;
  - Cognitive;
  - Language;
- Whether homework was assigned following the intake interview or during the assessment (including follow-up expectations);
- Whether AT was trialed or considered during a work or volunteer experiences, job shadows, or similar real-world vocational experiences (regardless of success);
- Availability of device reutilization or refurbishment;
- Specific AT devices or systems being recommended including
  - How the decision to select recommended items was reached;
  - How the AT objectively reflects the consumer's Great 8;
  - Discussion of consumer informed choice and how the consumer was involved in decision-making;
  - Cost for the recommended items (and cost-effectiveness/comparisons of the selected items as available and appropriate);
    - Single-source availability;
    - Customized or fabricated item;
    - Commercially available (off-the-shelf);
  - Comparable benefits (cost sharing or other funding sources);
  - Discussion regarding why the AT is reasonable, necessary, and appropriate;
  - Selected AT pros and cons;
  - Maintenance, repair, and warranty issues;
  - Availability of ongoing necessary AT services;
  - How soon, if known, the recommended AT devices will become outdated, obsolete, or be unavailable (is there a better long-term solution);
  - The use (efficacy) of each item in relation to all recommended items and any AT the consumer is already using;
- Has the rehabilitation team provided input into AT selection;
  - Disability-employment professionals;
  - Family;
  - Friends;
  - Educators;

- Medical professionals;
- Mental health professionals;
- Training considerations;
- The need and availability of wrap-around supports (e.g., personal cares, job coaching);
- Whether there is an appointment scheduled for report and plan review with the referral/funding source; and
- Implementation considerations.

An implementation (or installation) plan may be included in the report or it may follow once the report is approved and implementation funding authorized by the referral/funding source. The implementation plan typically consists of projections for completion dates for AT ordering; setup and configuration; and delivery, installation, and training.

REPORT ACCESSIBILITY

AT report content should be accessible to readers. AT reports should offer the greatest clarity for the greatest number of people. In other words, the intended audience should be able to readily understand it. The Center for Plain Language outlines four primary elements to make a document accessible:

- Findable: Content should be informative: it should state the purpose for which it was written, relevance for the reader, and any expectations the content places on the reader. It allows readers to easily locate the content for which they are looking;
- Readable: Content should be organized to allow for an easy flow. Write in an active voice, use familiar words, define unfamiliar terms, and avoid highly specialized terms, slang, and pejorative (biased or subjective) language;
- Understandable: Content should be written with the target audience in mind. The objectives for which the document was created should be clear and the content concise and unambiguous.
- Relevant: Content should be conveyed to the reader at a personal level. Personalization may involve summarizing the overall message, anticipating questions, offering examples, and considering next steps (Center for Plain Language, 2018).

Writing an AT report can be challenging because it is intended for very different audiences. Generally, this includes the referral/funding source and the consumer. However, with more than one funding source, and rehabilitation team members, there will be mix of educational levels and expertise. Plain language maximizes access to the report information for all readers.

CONSUMER RESPONSIBILITY

It's easy to single out the AT provider and their team members and take them to task over having the responsibility to move the AT service delivery process along in a timely manner. And they should be diligent in the performance of their duties. However, it is helpful to remember that the AT provider is typically not alone in providing AT services. There are other key players in the process, such as the disability-employment provider (e.g., vocational rehabilitation or independent living services) and the funding source (e.g., vocational rehabilitation, federal benefit, or insurance providers). They too bear some responsibility for the flow of timely services. We would also add the consumer of AT services to that list.

Consumers have a responsibility to be engaged in the AT service delivery process from start to finish—to the degree that they are capable. If they lack capacity for engagement, their representatives bear this responsibility. They should recognize (i.e., they should be directly informed at the outset of service delivery) that they will be asked to give full attention and effort and to persevere in the process (Duckworth, Peterson, Matthews, Kelly, & Carver, 2007). The more invested in the process they are, the more meaning and satisfaction they will derive from it. In disability-employment service delivery, the expectations placed on consumers are significant. If they are looking for full-time work, they should be available to engage in work-related activities for the same amount of time. In other words, if a consumer expects to obtain employment at 30 hours/week, they should have at least 30 hours/week for job development activities. Investment in the process is commensurate with the time they plan to invest following job hire. Engaging with rehabilitation team members to coordinate job development activities is included in this investment of time (e.g. assistive technology services).

SUMMARY

The AT report (the assessment summary, service plan, recommendation, and implementation plan) is developed from the information gathered during the assessment process. Consumer need areas, which were identified and clarified during the assessment, are matched with AT features and devices or systems. Recommendations are presented and discussed in the AT assessment report. Manufacture details are included and the expected cost of the technology is listed. AT progress measures should be structured for clarity and measurability so that consumers have a clear understanding of the expectations the AT professional or rehabilitation team have around AT service delivery, so progress can be monitored and the vocational goal can be achieved. The AT report should be accessible (i.e., written in plain language) to consumers and the rehabilitation team alike (e.g., readily understood regardless of educational background or professional expertise). Consumers have a critical responsibility (alongside the AT professional and members of the rehabilitation team) to invest themselves in the process. Time, effort, and persistence are necessary if recommendations are to be successful in addressing their vocational needs.

REFERENCES

Center for Plain Language. (2018). Five steps to plain language. Retrieved from ⟨http://centerforplainlanguage.org/learning-training/five-steps-plain-language/⟩.

Csikszentmihalyi, M. (2003). *Good business: Leadership, flow, and the making of meaning.* New York, NY: Penguin Books.

Duckworth, A., Peterson, C., Matthews, M., Kelly, D., & Carver, C. S. (2007). Grit: Perseverance and passion for long-term goals. *Journal of Personality and Social Psychology*, *92*(6), 10871101.

Haworth, J. T. (1997). *Work, leisure and well-being.* New York, NY: Routledge.

University of Maine System. (2018). Human resource: Setting goals together. Retrieved from ⟨http://www.maine.edu/pdf/pasetsmart.pdf⟩.

Problem-solving

9

Ray Grott

Rehabilitation Engineering and Assistive Technology (RET) Project, San Francisco State University (SFSU), San Francisco, CA, United States

PROBLEM-SOLVING FOR ASSISTIVE TECHNOLOGY SOLUTIONS

INTRODUCTION

In this chapter, we discuss how to arrive at successful recommendations and design of assistive technology tools for use by consumers with disabilities. Our goal will be to demystify the process of selecting or designing the most appropriate solutions that have the highest likelihood of meeting everybody's needs. While the focus will be on employment settings, the same approach can be used in the school, home, and community.

Various "models" of assistive technology service delivery have been identified and utilized over time. These include what Marcia Scherer and others have identified as the rehabilitation model, the needs-based model, the human activities and assistive technology model, and the matching person and technology model (De Jong, Scherer, & Rodger, 2007). Anthony Shay, in Part 1 of this text, offers a new framework he calls the accommodation system model that incorporates AT in the model's accommodations dimension. While different approaches and theoretical frameworks have their strengths and weaknesses, the AT practitioner is typically expected to come up with the best possible solution within the shortest amount of time—and usually at the lowest cost. For that, we will outline a time-proven "problem-solving methodology" that has been taught and utilized in various forms by engineers and designers for decades. When approached with an open and exploratory mind, this method can foster creative and unique approaches to solutions, which is the hallmark of our field.

THE IMPORTANCE OF SYSTEMATIC PROBLEM-SOLVING

The first questions people might ask are, "Why do I need to follow a methodology? Why can't I just go in my shop or meet with my carpenter friend and cook something up as I've done in the past?" In fact, this is what many home-grown inventors

Assistive Technology Service Delivery. DOI: https://doi.org/10.1016/B978-0-12-812979-1.00009-6

and "makers" do (Willkomm, 2013). There's nothing wrong with this—when it works. Failing is part of the learning process. Timeliness, however, is also a part of the service delivery process. The problem is that ad hoc solutions often *don't* work and meanwhile the prospective end-user is waiting (more or less patiently) for something that *does* work. Their ability to get or keep their job can depend on it. There is also the potential for frustration on the part of the end user or employer with solutions that don't work. This may in turn generate a reluctance to try other solutions or the decision to abandon the effort altogether.

A clearly defined problem-solving method serves a number of important purposes. Following a structured approach, asking the right questions, and gathering the right information can help avoid problems early on or later down the road as circumstances or user abilities change. It can help the AT service provider remain focused and stay on track. It can counteract the tendency to start working on problems that don't exist or may not need to exist. It can restrain the designer from going off on tangents exploring possible solutions that don't fit the reality of the user's needs, abilities, or environment. In addition, demonstrating that you are using a systematic approach can be reassuring to the consumer, employer, and funder who are depending on you. Finally—and most importantly—a clear methodology can maximize success and reduce the number of revisions or even failures.

TIP: JOB MATCHING

Assistive technologists and rehabilitation engineers, disability-employment professionals, and allied health professionals—working individually or as a team—are not miracle workers. We need to be prepared to acknowledge when it is not possible to adequately accommodate a consumer's functional abilities and therefore the job is just not a good match for the individual. A good, well-documented assessment process can help the various stakeholders better understand and accept this outcome and point the way to more suitable employment.

A PROBLEM-SOLVING METHODOLOGY

In the following sections, we will discuss the components of a problem-solving methodology that can be used by people with different backgrounds or training, whether or not they are a skilled rehabilitation engineer or AT practitioner. While some of the points may seem obvious to people with more experience, even the most seasoned among us make mistakes when we fail to consider some of these steps.

- Define the problem
- Analyze the problem
- Restate the problem and goals
- Establish criteria for the solution(s)
- Generate solution ideas

- Evaluate possible solutions and select the best solution(s)
- Implement solution(s)
- Assess, Revisit, refine, etc.

DEFINE THE PROBLEM

The path to a solution begins with clearly identifying the problem that needs to be solved. Defining the problem accurately and specifically is therefore a critical step in the problem-solving process. Mistakes here can lead to wasted effort and frustration. Quite often, the problem that is presented to us in referrals or requests for assistance is not clear, focused, or accurate. Sometimes the rehabilitation counselor, manager, or consumer is working from limited information or is influenced by prior situations or suggestions by others. For instance, a request can come in the form of a specific solution: "The consumer probably needs an ergonomic keyboard" or "I need a trackball like my coworker has." Sometimes a referral can be overly general and indicate only that "The consumer hurts when sitting."

It is important to note that the first statement of the problem is often not the core or even the correct one. Once more research is conducted, we can more clearly define the problem, while deferring the solution until later in the process. At this early stage, it is important to avoid making quick assumptions, jumping to conclusions, or assuming that other stakeholders have adequate information. By using the above example, "needs an ergonomic keyboard" might more accurately become defined as "needs a way to work more efficiently and with less pain, despite the consumer's carpal tunnel syndrome." Even this more encompassing statement of the problem may not prove adequate. A bit further along in the process, it might be modified to add "primarily when working with bookkeeping software on a notebook computer at clients' offices." Often a consultation to address one stated problem uncovers multiple issues that require attention.

Here are some other examples of the difference between the problem as stated and the problem as more clearly identified later:

- Someone suggests the provision of "screen magnification software." An analysis reveals that the person needs a way to "make objects on the computer screen 1.5 times larger." Depending on a number of factors, the recommended solution might range from using settings built into the operating system, web browsers, and applications, to providing a larger monitor, to using screen magnification software.
- A bookkeeper requests that someone drive her to the bank to make deposits. After exploring the situation, it is clarified that she has a neck injury and experiences sharp pain when rotating her head to safely change lanes or back into a parking space. While a wide-angle rearview mirror is recommended to resolve this particular issue, other problems relating to the need to turn her head are explored. This reveals concerns regarding the positioning of printed documents to the side of her computer monitor and how she views the off-center monitor.
- A sit/stand workstation is requested to accommodate someone with back pain while sitting. A more detailed exploration clarifies that the person cannot tolerate standing in a fixed position in front of a desk for even short periods of

time. The challenge is redefined as devising a work station where the consumer can alternate between working while sitting upright in an adjustable chair and working in a significantly more reclined position to off-load weight and pressure on the spine.

NOTE: The discussion that follows is framed as if the person needing the assistive technology supports already has a job or is hoping to return to an existing job after losing some functional capacity. In many cases, the AT provider may be asked to work with someone who is seeking employment, often for the first time. The request can be to help the person and their support staff, such as vocational rehabilitation counselors or teachers, understand what is even *possible*, given the appropriate AT. (This can be particularly important for people who have lost function due to an injury or medical event and think their working life is over. People who have grown up with a disability will hopefully have had some introduction to AT, though that is certainly not always the case.) Once an employment goal has been identified, the next step would be to help them identify, acquire, and train on the technology tools they will need to use once they get a job. Often the employment goal will require further education or vocational training, in which case the same or additional AT may be needed for that stage of the employment process.

ANALYZE THE PROBLEM

Once the problem (or problems) is more clearly defined, we can begin to ask questions and do some research into the various elements surrounding it. Depending on the complexity of the situation, this can be a short process undertaken during a 1-hour assessment, or it can be a protracted effort spanning weeks. The basic question we are asking is "what do we need to understand better" about all the key aspects of the problem. Here are some of them.

Requirements of the job

When conducting a worksite accommodation, it is important to understand the actual job tasks and the essential functions of the job. One can quickly discover that there is a significant discrepancy between the job description and what a consumer really does. Some employers may provide only brief and vague job descriptions, while others go into greater detail about the tasks and expectations. It often falls to the person hired to recommend accommodations, including technology-based ones, to conduct a job analysis. This should focus on the purpose of the job and the importance of the job functions in achieving that purpose. In the United States, the Americans with Disabilities Act (ADA) talks about "essential functions." Briefly, these are the activities that the job position was created to perform—and these are the ones that should demand the most attention from service providers.

For instance, in the example above, one might ask to what extent the book-keeper was hired to keep the accounts versus travel to the bank to make deposits. One could determine that keeping the accounts is an "essential" function, while taking deposits to the bank is a more marginal one (i.e., tasks not essential to the function of the position). It would be a real problem if the bookkeeper suffered a brain injury which resulted in an inability to manage numbers. On the other hand, a truck driver tasked with making deliveries is expected to transport materials by personally driving a truck; the purpose of driving the truck is to make deliveries and is an essential function of the job. Additionally, some jobs have expanded or contracted from their original description based on what a prior consumer was inclined to do or was particularly skilled at. Perhaps the previous bookkeeper liked doing graphic design and volunteered to produce the company's monthly employee bulletin. That doesn't mean that the next person filling this role should be expected to continue doing this, unless the job description is modified to include that skill.

One way to help understand which job functions are essential is by breaking them down into the components: Action, Object, Function, and Purpose. For our bookkeeper, the "Action" would be to drive, the "Object" would be a car, and the "Function" would be to transport herself to the bank for the "Purpose" of making a deposit. In this instance, one may need to ask whether it is essential for her to personally get to the bank to make deposits before addressing how to help her drive. And even then, there may be alternative ways for her to get to the bank, such as taking the bus or taxi or getting a ride from another consumer. For the truck driver, the "Action" would be still to drive, the "Object" would be a truck, and the "Function" would be to drive the truck for the "Purpose" of delivering materials. Clearly, the materials could not be delivered if the driver was not able to drive.

In addition to considering whether the position exists primarily to perform the function that the person was hired for, other determinants of whether a function can be considered essential include the amount of time spent performing the function, the consequences of not requiring a person in this job to perform the function, the experiences of people who have or who are currently performing similar jobs, and even the employer's judgment about what is essential. This can be a contentious issue when considering accommodations and information resources especially when individual rights are governed by legislation. For our purposes, understanding what tasks and activities are essential versus marginal can help the technology consultant establish priorities and focus on an individual's core needs.

Additional core components of a job analysis are understanding the usual and customary methods used to perform the job tasks, the machines and tools typically utilized, and environmental conditions such as noise, temperature, and possible hazards. On the human end, one needs to understand the physical demands placed on the consumer, the focus and concentration required, and their expected educational level and intellectual capacity.

The evaluator will also want to gather information on the performance standards and expectations for the consumer's position and obtain a general idea of whether the consumer is meeting, exceeding, or failing to meet them. If this is a "job retention" situation, and the person is in danger of losing the job, it will be important to understand how much of an improvement is being sought or required and in what the expectation is for how long this should take.

While more difficult to assess, some attention should be paid to the resources available to the employer. Is this a large enterprise with "deep pockets," a small business with a few employees, or a struggling nonprofit agency? This might impact whether the employer might feel that the costs of a recommended technology-based accommodation is an "undue hardship" (in the language of the ADA) and be reluctant to pay for it. In that case, another funding source such as a vocational rehabilitation agency may be able to help with the equipment costs.

The consumer's abilities and functional limitations

An accommodation or technology solution succeeds or fails based on an accurate understanding of the individual who will be using it. It is therefore important to explore the person's abilities, limitations, and preferences. As with vague job descriptions, general diagnoses or medical labels will not be sufficient for our purposes. Experience quickly teaches that no two people, no two bodies, are the same, even if they share the same medical diagnosis or physical limitations. Learning more about the individual nuances of the person we are working with can help us stay on track with our suggestions and recommendations. Having said that, the evaluator does not have to—and should not—pry into all the aspects of the person's disability and how it impacts his or her life. The focus should remain on the abilities, limitations, and preferences related to specific job tasks and work requirements.

With all due respect to the medical profession, often a doctor's recommendation or list of work restrictions is a "guesstimate" and many times is presented as a list of how much a person could or should not lift or carry, bend or climb, or how long they should perform an activity such as typing or using a mouse without a break, or a maximum total time that the task should be performed in the course of a work-day. This is typically based on a medical evaluation without an understanding of the specifics of the job or work environment. Even clinics staffed with occupational and physical therapists that perform work capacity evaluations can at best only simulate standard work activities without knowledge of the real-world workplace environment where these activities will be conducted. In addition, these profes-sionals may not be aware of possible AT solutions that can mitigate the physical work load and increase the consumer's function. While medical reports can be very helpful and should be read in advance of the evaluation, equally relevant informa-tion can be acquired by talking with the consumer (e.g., if the person does not yet have a job, one can ask about prior work, school, or home-based experiences). Finding effective solutions requires action and data collection (Costello, 2011).

Gathering information from an individual—spending sufficient time with them to accurately determine their needs—is an important part of problem-solving

(Maykut & Morehouse, 1994). The following are some of the questions that we might ask in a first meeting while trying to get as much detail as possible:

- What tasks have you determined you are able to perform? Which ones seem slow, difficult, or impossible?
- Can you easily reach or lift the items you need to work with?
- Do you experience pain? Where? During and after what activities? How quickly does it begin hurting?
- Does your energy and productivity vary during the day, or from day to day? If so, describe how it varies for me.
- How is your eyesight? Hearing? Speech consistency? (These issues are often overlooked.)
- Do you expect your disability or functional limitations to improve or get worse over time?
- Have you in the past or are you currently using any assistive devices or other accommodations?
- How is your memory? Do you have any difficulties with reading, writing, learning new material, or keeping track of tasks and responsibilities?
- Are you experiencing stress? How much (is it excessive, or a concern)? In what situations?
- Do you feel that others understand or are supportive of your limitations?

TIPS: GATHERING INFORMATION

1. It can be helpful to ask the same questions from different angles or frame them in different ways. For instance, a person may be in a bit of denial or overly optimistic about her or his physical abilities. In addition, people with disabilities often have a great capacity to "make do" and accommodate themselves to the situation, rather than having an expectation that the physical environment can be changed to accommodate them. Therefore, they may tend to say that they are doing well enough or that any given aspect of their situation is "OK" or "not a big deal." A common response is that they have just adapted to the situation. This needs to be acknowledged before suggesting that they might benefit from improvements. (Two simple and common examples are a nondisabled left-handed person moving the mouse to their left side but keeping their hand in an awkward position since they did not know that they can swap the mouse's button assignments. Or a one-finger typist constantly turning Caps Lock on and off to generate a capital letter since they had not learned that the Sticky Keys tool allows them to press the Shift and letter key sequentially.)

2. Throughout the process, the evaluator needs to be prepared to *learn from the consumer* about what they *can* do, and how they do it, despite preconceptions attached to their diagnostic label or how they appear

physically. Often people with disabilities come up with efficient compensatory strategies to complete tasks, which can be useful to learn and understand during the evaluation.

3. Establishing good rapport and a level of trust is important for getting honest and complete answers to questions that can sometimes trigger emotions or relate to bodily functions. There are many resources available for learning "active listening" and other communication skills.

One particularly challenging area of exploration is cognitive limitations that can impact a person's performance. These can stem from brain injury, chemical exposure, stroke, emotional trauma or stress, various medical conditions, medications, attention deficits, distractibility, or below average abilities. They will of course vary greatly between individuals. Many people may be aware of their strengths and weakness in this area, while others are not. While ideally a neuropsychiatric evaluation can be provided by the referring agency when appropriate, this is often not available. In any event, careful questioning, including a review of the person's past work history or school experiences, examination of current work output, and observation of the person performing their job tasks can be very informative. There are many tools designed for people with learning difficulties or disabilities that can help a broad range of people with reading, writing, organization, and time management. People with emotional or psychological problems can often benefit from tools that can help them focus on their task, while modifications to the work environment can reduce noise, distractions, and other stressors.

Observation and measurement

At some point in the problem analysis process, we need to get out the tape measure and other recording tools and closely examine the work area and the environment, observing the consumer performing the actual job tasks. Understanding the fit or mismatch between a person and their work environment and helping resolve concerns is the core of ergonomics and fits within the realm of human factors engineering and design, as well as rehabilitation engineering and assistive technology. While the "fit" between the person, task, and environment are important to everyone, they can be critical for someone with difficulty using one or more parts of their body (Taveira & Smith, 2012). Measurement data should be collected when applicable (Steinfeld, Lenker, & Pacquet, 2002) and may include:

- Anthropometric: related to the size and proportion of the body and its parts
- Kinematic: related to motion of the body and its parts without consideration of the forces exerted upon them
- Biomechanical: related to the force of muscle exertion and gravity on the body
- Environmental: related to physical factors outside of the person

The types of information listed below are key when identifying problem areas and considering possible solutions and alternatives. Equally important, having accurate information is critical to the successful implementation of a selected solution.

Measurements of the person and work station/tools

- Basic measurements of the individual's body (overall height, elbow height, eye height, knee to heel height, reach range, etc.)
- How high is the work surface or machine?
- How much higher or lower should it be to match the consumer's comfortable and functional working height?
- How far in any direction is the consumer currently required to reach?
- How far can the person reach without discomfort or significant effort? How long can the person reach without discomfort or significant effort?
- How much push, pull, turning, or grip is required?
- How far is the visual target, such as a computer monitor? What is the size of the visual target, such as the text on the monitor screen?
- What is the person's comfortable seating height (in an office chair, stool, or wheelchair)?
- How do visual and manual working heights vary between sitting and standing?
- What are the individual's needs and preferences (such as sitting versus standing, right versus left hand use, orientation in the available work space, lighting and temperature levels)?

The work environment

- What environmental factors are noted? (Noise levels, light levels, temperature, odors, distractions.)
- Is more than one person using the same work station? (Need to consider their needs and openness to change. Ergonomic principles should be considered for these users as well.)
- General access for wheelchair or scooter riders, people using canes, crutches, or having other mobility needs.
- Path of travel. (Does someone with sensory loss need to navigate a complex or potentially dangerous path, such as a warehouse with many forklifts moving about?)
- Proximity to others? (Possible distractions to focusing on job tasks, limited privacy for using speaker phones, or speech recognition software.)

Other aspects that may be less easily quantifiable but are also important to note

- Job task sequences: are they logical and efficient?
- Time-based tasks: does the pace established seem reasonable or achievable?
- Task complexity: are they intuitive and reasonable for the person?
- Is training available?

- Are training materials available in formats suitable for the consumer?
- Apparent support, or lack thereof, from employer, supervisor, or coworkers?
- Personal preferences (not to be ignored or discounted!): including orientation within the available work space, overall appearances, the color of possible accessories or technology aids?

Other aspects related to the person that should be considered:

- Anticipated changes—will the level of functioning decline or improve over time?
- Will skills improve through practice and training?
- Has adequate training been provided?
- Are there competing priorities: motivation is ostensibly to obtain or keep a job but motivation may shift (or remain focused on) maintaining public or private income supports (e.g., disability insurance or social welfare, etc.).

CASE STUDY: THE WAREHOUSE EMPLOYEE

The degree of support and buy-in from management and supervisors can be critical to the successful employment of a person with a disability. For example, a deaf consumer with a large distribution company applied for a transfer to the warehouse operations as a stock picker. Management was concerned about safety issues and brought in an AT specialist. Issues raised by the employer included being aware of the movement of forklifts, hearing emergency alarms, communicating with supervisors and coworkers in a fast-paced environment, and participating in training sessions. After a detailed exploration of the work environment and close questioning of the key parties, the AT specialist participated in a meeting with management, the consumer, and his union representative. The specialist provided point-by-point scenarios as to how management concerns could be addressed, and the union representative gave examples of similar worksites where Deaf workers were successfully employed. They suggested that a supportive and collaborative environment could be fostered where the consumer and his coworkers could problem-solve minor issues as they arose. On the management end, the direct supervisor and his superior continued to raise concerns that appeared to be based on a lack of understanding of how people with hearing loss function in the broader world and continued to concentrate on why this situation might not work, vs how it could work. Ultimately, the consumer gave up the effort and secured a similar warehouse job at a different branch of the company that was more supportive.

RESTATE THE PROBLEM AND ESTABLISH CRITERIA

Once the areas of "mismatch" between the person and the task, tools, or environment are identified, we can more accurately identify the problem or problems that need to be addressed. A proper problem statement effectively captures the situation and allows us to better understand the goals (Morehouse, 2012). For instance, we can move beyond "the consumer experiences pain" to "the consumer needs to approach and utilize her work surface which is positioned at an appropriate height which allows greater comfort and which is designed to bring her work tools into closer proximity to her addressing access needs due to her weak arms and limited reach range." Or, from "he needs a foot-controlled mouse" to "he needs a way to activate a left mouse click without using his right index finger."

Establish Criteria for the Solution

It is important to establish clear goals for each problem we decide to address, going beyond the Problem Statement to include measurable criteria that will guide the design. The more detailed and specific the solution criteria, the less likely a key element will be missed in the process of selecting and implementing a solution.

For our consumer with the workstation problem, we could state "the consumer needs an adjustable-height stool that can bring her elbows up to 30″ high, as well as a work surface with a cut-out designed to bring her tools within 8″ of her torso due to her weak arms and limited reach range."

Prioritize the problems

Once we identify the problems, we need to prioritize them to determine which problems are more pressing relative to the others. This should have already been done to some extent in our discussions with the consumer and supervisor. It relates directly to what percentage of the job is impacted by each problem, how readily it can be addressed by other means (such as having someone else do a task), how "essential" it is to the person's core functions, and how immediate the concern is. Of course, some of the problems can be addressed simultaneously. Others may require more protracted efforts to solve and need to wait until the "easier" ones are implemented.

GENERATE SOLUTIONS

People with experience in assistive technology or just problem-solving in general understand that there are multiple ways to address a design challenge. People draw on their personal skills, materials preferences, prior work, and flashes of inspiration to arrive at proposed solutions. That said, not all solutions are equally appropriate or effective. This is where a systematic process can lead to more successful outcomes.

There is of course a big difference between problems that require a relatively simple solution (for instance, a table at a specific height or a compact keyboard) and those that may need some novel technology applications. Nevertheless, there are still considerations for selecting the most appropriate table or keyboard. In the rest of this section, I will assume that a more complicated situation is being addressed.

TIP: AVOID MAKING ASSUMPTIONS

Don't take what exists for granted. Often a physical obstacle or inefficient technique stays in place for years because everyone assumes someone put it there or developed it for a good reason, but no one remembers who initiated it or why. The work environment, method, or assignment can often be changed.

For example, a job accommodations expert was working to help place a woman with one functional arm into a custodial position. He and his team successfully addressed many of her job tasks but were concerned about how she might be able to tie knots in the large trash bags before tossing them in the dumpster. Fortunately, they asked her supervisor about the importance of this step and were told that there was no particular reason for tying a knot—it was just how others did it. They agreed that twisting the top together to keep the contents in the bag while being deposited in the dumpster would be adequate. Through observation, reflection, planning, and action, the problem was resolved (Costello, 2011). This was a much simpler solution requiring minimal time and effort.

Review initial ideas

It is quite common for the evaluator to begin generating ideas for possible solutions early in the problem-solving process—often before the information gathering is complete. Often the consumer (employee), the supervisor, or a technically minded friend or coworker can be a good source of initial ideas. It is best to note these ideas and then put them aside for the time being while focusing on getting adequate and accurate information. Once we enter the solution generating stage, it becomes time to revisit this list, discarding ones that are no longer appropriate and bringing forward those that appear to have potential.

Explore and research possible existing solutions

The first question to ask is whether a solution already exists. A common mistake inventors make—especially home-grown designers and students—is to essentially reinvent the wheel. While there are not always good avenues for people to share their ideas, the growth of the Internet has made this a lot easier—although it is still a scattered process. There are a few websites specifically for sharing AT designs, such as AT Solutions (www.atsolutions.org). There are lots of generic or disability-specific adaptations listed on YouTube and various "maker" sites such as Instructables.com, though that can be a time-wasting expedition unless one is disciplined and focuses on the problem at hand. It might be more productive to search disability-related sources of existing products such as AbleData.com and Assistivetech.net. Searches through equipment supply sites such as Global Industrial or Grainger, reseller sites such as Amazon, and company-specific sites can uncover possible products that can be used as-is or modified, or that at least can help generate some ideas.

The availability of Internet searches shouldn't keep us from drawing on one of our most important resources, which is the knowledge, experience, and willingness to share information that is characteristic of assistive technology professionals. Options range from contacting AT providers directly, utilizing listservs such as the AT Forum sponsored by RESNA (The Rehabilitation Engineering and Assistive Technology Society of North America), the Assistive Technology Professionals group on LinkedIn, and disability-specific websites. In Europe, there is the European search engine on assistive technology put together by EASTIN (European Assistive Technology Information Network, www.eastin.eu).

Although these peer-based networks can be excellent sources of ideas, it is important to be aware of organizations staffed by experienced people whose job it is to help people find the tools they need. While in the United States, most states receive funds to provide information and referral advice though the federal "Tech Act," the most important resource from the perspective of this book is the Job Accommodation Network (JAN, https://askjan.org/). JAN employs advisors with expertise in different disabilities and branches of AT who can offer suggestions of physical as well as more administrative accommodations. However, it is important to note that they are working remotely with limited information and their ideas should not take the place of a more detailed in-person evaluation.

Brainstorming

Another common method for generating ideas is "brainstorming," where the general goal is to generate many ideas of all types, without any judgment or critique. This can involve multiple people or just one person writing down a list of key words or phrases that come to mind. While the "right" solution doesn't automatically surface during this process, brainstormed ideas can frequently lead to insights and linkages that end up being useful. I find that just browsing through a hardware store or catalog with an "open mind" can help lubricate the creative process. For those who need some help getting started, there are many books available on brainstorming and fostering creativity and invention.

Problem elimination

One important solution that should always be considered is not having the consumer do the task to begin with. This might be accomplished by restructuring the job so that someone else takes on the task. Drawing on our earlier example, perhaps someone else can deliver the bank deposits other than the bookkeeper. Or maybe this can be done electronically using new banking tools rather than driving to the bank. ("Making bank deposits" does not necessarily mean that this activity must be done in person at the bank.) A corollary of job restructuring is often job sharing, whereby two employees might swap tasks. If someone else is asked to make the bank deposits for the bookkeeper, the bookkeeper might perform a task that the other employee typically does, such as ordering the office supplies. As implied elsewhere, accommodating one employee should not create a burden on others. Equitability and comparable job expectations—along with a flexible and supportive work culture—can help avoid feelings of resentment or privilege, which can undermine other efforts at implementing accommodations.

TECHNOLOGY SOLUTIONS ARE NOT THE ONLY ONES

As discussed in Part 1 of this text, the provision of AT is only one of a variety of ways that a consumer's needs can be addressed, and it is important to put the technology-based solution within the context of other equally valuable and important accommodations. In relation to the ADA, "reasonable accommodation" is any modification or adjustment to a job or the work environment that will enable a qualified applicant or consumer with a disability to participate in the application process or to perform essential job functions (ADA.gov, 2018). Any website or printed material on possible accommodations will list options such as job restructuring, modifying work schedules, providing personal assistance services (such as sign language interpreters, readers, or scribes), and even job reassignment to an available open position within the company. While most of these might be considered administrative solutions that entail no or limited costs, personal assistants can be costly, but their need can be reduced in many situations using standard or specialized technologies.

For example, Paul worked as a transcriptionist for a government agency. He developed cellulitis in his legs and feet, making it hard for him to use the traditional foot pedals that controlled the playback functions on the transcription machine. During my problem analysis, I noted that his chair was too small for his large frame, forcing him to place more weight on his feet. I also learned that he had a difficult time walking down the hall to the central printer where he would retrieve the results of his work. Additionally, his doctor had recommended that he periodically elevate his legs to help reduce their swelling, but he had no time or location to do this. The "technology" part of the solution involved tapping into the wiring of the transcription control box and mounting flexible switches adjacent to his knees to activate the two most-used playback functions. The other equipment component was providing him with a more suitable chair. However, equally important to the accommodation package, was relocating his office cubicle closer to the printer and providing a simple couch in the (large) men's bathroom, along with longer break times that he could deduct from his lunch hour.

EVALUATE AND SELECT SOLUTIONS

At some point, the individual or design team needs to pick the best and most feasible solution and start working towards its implementation. While it can be clear which possible solution rises to the top, the process sometimes requires both a critical analysis of the competing ideas and a way to rank them. Ranking criteria might include the extent that a candidate meets all the design criteria, ease of implementation, time to implement, cost, etc. Before assigning a score to each of these criteria, they can be given a "weight" of relative importance to help one end up with a leading contender. There are many examples of Decision Matrices on the Internet.

1. The service provider has a conversation with the consumer and the rehabilitation team (including the employer) as appropriate and winnowing away those criteria which do not reflect an immediate need.
2. Based on purpose and relative importance, each criterion is weighted to reflect consumer need.
3. Create a table with decisions for row headings and criteria for column headings. List their assigned weights (importance ratings).
4. The service provider then analyzes the decisions while considering the criteria. Rating scales or rank-orders may be used to determine the goodness-of-fit of a proposed solution.
5. For each criterion, multiply each idea by its importance rating (moving horizontally across each row). Add the points for each option (moving vertically summing the numbers at the bottom of each column). Options with the highest scores will not necessarily be best choices but do provide information toward meaningful discussions around determining effective and satisfying AT.

USING A DECISION MATRIX

A decision matrix can offer a means to discuss options with consumers and the rehabilitation team. It can supplement a report provided to a referral/funding source to facilitate planning and funding decisions. It may also help frame the decision-making process for them in terms which allow for consumer buy-in regarding effective AT choices when they have a difficult time selecting an item which effectively matches their needs (ASQ, 2018). Fig. 9.1 provides an example of a decision matrix.

Whether service providers use a formal rating system, discuss the options with others, or think it through on their own, there are key factors that need to be kept in mind. As mentioned, meeting the main design criteria is central, while ease of implementation, cost, and time to deliver can be determining factors. Other important considerations include

Decision / Criteria	Importance rating*	Idea A**	Idea A score***	Idea B	Idea B score	Idea C	Idea C score
Addresses need	4	3	12	4	16	3	12
Safe	4	4	16	4	16	3	12
Simple design	3	4	12	3	9	2	6
Speed of implementation	3	3	9	3	9	2	8
Reasonable cost/affordable	2	4	8	3	6	2	4
Consumer likes	3	3	9	4	12	3	9
Aesthetics	2	2	4	3	6	3	6
Weighted totals			70		74		57

*Importance rating: 4 = required, 3 = very important, 2 = somewhat important, 1 = desirable.
**How well the criteria are met Criteria: 4 – 1, with 4 being the best.
***Score = importance rating x how well idea meets criteria

FIGURE 9.1

Example of an AT decision matrix.

- The solution's possible impact on other aspects of the job, including traditional methods and procedures, work flow, task allocation, etc.
- The need to change the environment, such as the size of the workspace, relocation of workspace, modification of lighting or ventilation.
- The solution's possible impact on other consumers. (For instance, would using speech recognition or speech output software impact other people nearby and could that be easily mitigated? Are others using the same workstation or tools that might be modified?)
- The required support and buy-in from supervisors and coworkers.
- The need for training and practice, which can take away from time devoted to work tasks.
- Are there union rules or legal constraints that could complicate implementation?
- The availability of resources needed to carry out the solution.

FINAL DECISION MAKING

When it comes time to decide which solution(s) to pursue, a combination of factors come into play. The result of any formal evaluation matrix should be carefully considered. Often the results of the formal or informal decision process need to be discussed with other stakeholders who may not have been part of that effort. The consumer must be consulted for their opinion, ideas, concerns, etc.

While legislation may not mandate an employer to provide the specific accommodation requested by an employee, consultation and discussion at this stage can avoid disagreement, disappointment, resentment, and even rejection of the accommodation and related AT tools. This is discussed in more detail in the chapter on implementation. The employer or supervisor must be consulted, as they ultimately oversee the consumer, evaluate their performance, and control the work environment—even if another agency is funding the accommodation.

Additional information may be required before proceeding to final implementation. For instance, a new workstation setup may need to be mocked up and tested in actual use—as opposed to a quick simulation during the evaluation. At other times, different hardware or software may need to be loaned or installed on a trial basis to see which one best meets expectations. In some cases, subjective feedback is adequate, while at other times objective measurements of relative output and productivity may be called for. Input from the user needs to be solicited, understood, and factored in before moving forward.

The above is the "ideal" textbook version of the problem-solving method related to job accommodations and AT. Real-world, time-driven contingencies often require quick decisions to be made following a relatively informal process. The success of any truncated approach is influenced by the experience of the provider, adherence to good accommodation principles, and attention to detail.

SUMMARY

The AT problem-solving process requires a consumer-driven approach to determining the most effective solutions, in a timely manner, and as cost effective as possible—a challenge that professionals working in AT service delivery typically meet with creativity, flexibility, and innovation. A systematic problem-solving methodology facilitates clarity of purpose, helps determine the right questions to ask, and ensures effective information gathering that forestalls future problems and reinforces a sense of resiliency. Effective problem-solving essentially keeps us focused on the actual problems, helps us make sense of the choices we are presented with, and guides our decisions regarding the best course of action to meet consumer needs.

REFERENCES

ADA.gov. (2018). *Americans with Disabilities Act: A guide for people with disabilities seeking employment*. Retrieved from < https://www.ada.gov/workta.htm >.

ASQ. (2018). Learn about quality: Decision matrix. http://asq.org/learn-about-quality/decision-making-tools/overview/decision-matrix.html.

Costello, P. J. M. (2011). *Effective action research: Developing reflective thinking and practice* (2nd Ed.). London: Continuum.

De Jong, D., Scherer, M. J., & Rodger, S. (2007). *Assistive technology in the workplace.* St. Louis, MO: Mosby.

Maykut, P., & Morehouse, R. E. (1994). *Beginning qualitative research: A philosophic and practical guide.* London: The Falmer Press.

Morehouse, R. E. (2012). *Beginning interpretive inquiry: A step-by-step approach to research and evaluation.* New York, NY: Routledge.

Steinfeld, E., Lenker, J., & Paquet, V. (2002). *The anthropometrics of disability: An international workshop.* University at Buffalo, Center for Inclusive Design and Environmental Access. Retrieved from http://idea.ap.buffalo.edu/Anthro/The%20Anthropometrics%20of%20Disability.pdf.

Taveira, A. D., & Smith, M. J. (2012). Social and organizational foundations of ergonomics. In G. Salvendy (Ed.), *Handbook of human factors and ergonomics* (4th Ed., pp. 274–297). Hoboken, NJ: John Wiley & Sons, Inc.

Willkomm, T. (2013). Assistive technology solutions in minutes: Ordinary items extraordinary solutions. Durham, NH: Institute on Disability, University of New Hampshire.

Assistive technology design and fabrication

10

Ray Grott

Rehabilitation Engineering and Assistive Technology (RET) Project, San Francisco State University (SFSU), San Francisco, CA, United States

In Chapter 9 on Problem-Solving, we discussed how an appropriate employment accommodation can be arrived at using a comprehensive problem-solving methodology. In this section, we will explore concepts related to the actual implementation of technology-based solutions. We will review some general concepts related to design, and discuss the implementation process. (We will talk about how solutions can miss their mark in Chapter 15: On technology abandonment or discontinuance.)

A SPECIAL NOTE TO THOSE WHO ARE NOT DESIGNERS OR FABRICATORS

The authors have written this book for people in a range of professions, including disability-employment professionals, assistive technology specialists, employment specialists, job developers, job coaches, technology users, and people with technical skills who have been asked to assist any of them (in effect, rehabilitation team members or wrap-around service providers). We don't expect that everyone has the interest and ability to do custom design and fabrication work. Nevertheless, it will be helpful to understand the context within which successful solutions are developed. This can enable the various stakeholders to identify the best resources to use and to help the designer "stay on track" with regard to the previously identified goals and solution options. We therefore encourage you to read the next sections—we promise you that we will not be discussing anything too technical.

THE HIERARCHY OF SOLUTIONS

Once we have clarified and refined the problem and come up with preferred solutions using the process laid out in the previous section, there are a few more aspects to consider before we reach for a screwdriver or turn on the table saw.

Assistive Technology Service Delivery. DOI: https://doi.org/10.1016/B978-0-12-812979-1.00010-2

There is a basic "hierarchy" of solutions that experienced assistive technologists and rehabilitation engineers pay attention to. The options range from basic procurement (i.e., purchase of technology) to creation of a technological solution. These are discussed below, with the most preferred options listed first.

- Off-the-shelf product designed for general use;
- Off-the-shelf product designed specifically for use by people with physical, sensory, or cognitive limitations;
- Off-the-shelf product that can be modified; and
- Custom designed and fabricated device or system.

OFF-THE-SHELF PRODUCT DESIGNED FOR GENERAL USE

Items in this category would be the first choice for several reasons. They are more likely to be produced in larger quantities and therefore cost less. Products will often have had durability and reliability testing before mass production. Warranty repairs and replacement parts may be available to reduce downtime if something breaks. A larger user base can make it easier to find replacements or parts after the product has been discontinued. On the administrative side, the procurement process can be simpler and faster for an off-the-shelf item. Finally, a device that looks like a standard consumer product may be more acceptable to the end-user and generate less attention and stigma in the eyes of others.

OFF-THE-SHELF PRODUCT DESIGNED SPECIFICALLY FOR USE BY PEOPLE WITH PHYSICAL, SENSORY, OR COGNITIVE LIMITATIONS

Compared to general-use products, those serving a specialty market can be expected to be more expensive due to higher development costs and lower sales volume. Often these products are produced by smaller companies that can come and go more frequently, along with their inventory. At the same time, it is important to acknowledge and support the major contributions that disability-oriented companies have made to the availability of assistive technology tools for thousands of people.

A valuable bridge is emerging between these first two categories as companies and designers are understanding the value of universal design (UD) as they offer products that are more usable by a wider group of people, especially as the population ages. A good example of this is the emerging availability of kitchen utensils, small appliances, and personal care and garden implements that are promoted as being easier to use by people with arthritis and similar hand problems. Additionally, standard software products are increasingly adding accessibility features, requiring less reliance on specialized tools. (See discussion of "Universal Design" later in this section.)

OFF-THE-SHELF PRODUCT THAT CAN BE MODIFIED

We can often identify a product that comes close to meeting the user's need but requires some additional modification. For instance, someone with unusually short arms may need a keyboard tray that comes up higher or can be positioned at a more extreme angle than is available. Having the basic tray mechanism can be a great start onto which a fabricator can add additional parts to attain the desired height, angle, and tray size. Another example would be a specialized mouthstick or hand typing aid that is not the desired length and may need to be cut down or extended in the shop. The clear advantage of this approach is that a lot of the design and fabrication has already been done, resulting in lower costs. Additionally, replacing or duplicating the device, or modifying it as the user's needs and abilities change, is generally cheaper and can be implemented more quickly.

CUSTOM DESIGNED AND FABRICATED DEVICE OR SYSTEM

This may come as a disappointment to readers who want to design and build items for people with disabilities, but custom-designed solutions are the last option in our hierarchy of solutions. While starting from scratch can be more interesting and challenging, it is almost always more time-consuming and therefore cost more to implement. On the plus side, the custom design isn't limited by the compromises often required when using a commercially available device as a starting point for a custom modification. Custom design can address novel and complex needs and the item can be tailored to the user's exact specifications. However, repairs and replacements can also be harder to achieve, especially by others not involved in the original effort. A lack of access to people with the necessary talents and skills for custom design and fabrication (along with needed tweaks or repairs) can make it difficult to implement the desired solution. Finally, possible liability issues can arise when something goes wrong and it is advisable that the fabricator has adequate insurance if charging for this service.

As a counterpoint to the arguments made above, it is important to note that "low cost" is a relative concept. One must consider not just the cost of the device or system, but its durability, reliability, reparability, and replaceability. Initial cost is less important in the long run than what is appropriate and works best for the end-user, especially if it helps them obtain, maintain, or be more efficient in their job.

DESIGNING CUSTOM TECHNOLOGY SOLUTIONS

Purchasing and installing off-the-shelf products can be relatively straightforward when done within the guidelines of the party that is paying for them. (Properly adjusting and configuring them will be discussed later.) Custom modify an item or designing and fabricating one from scratch involves a larger set of considerations. It is tempting to take the general design specifications identified in the problem-solving process and head into the shop. However, to be a successful rehabilitation engineer or assistive technology (AT) specialist one must consider some important design concepts in the development process.

USER-CENTERED DESIGN

The concept of user-centered design (UCD) has been around for a long time but is gaining more traction in recent years, especially as "usability" considerations have become more important in software and web design. Rehabilitation engineers and assistive technologists might claim that their work is, by definition, user-centered as it is aimed at enhancing an individual's function and well-being. However, that does not ensure that this extends into the design process. UCD is a person-centered design process in which user requirements, goals, and tasks are considered as early as possible. Early consideration lends itself to greater design flexibility and the likelihood of greater cost-effective changes (Czaja, & Nair, 2012, p. 50). Similarly, usability.gov, which offers a wide variety of resources on the topic, suggests: "The user-centered design process outlines the phases throughout a design and development life-cycle all while focusing on gaining a deep understanding of who will be using the product" (2018). While there are other definitions and variations of the UCD process, they share some core principles (Fig. 10.1):

UCD principles: UCD is characterized as a multistage problem-solving process that requires designers to

- Analyze and foresee how users **are likely to use a product**;
- Test the validity of their assumptions **in real-world tests with actual users**;
- Test at each stage of the design and production process creating a circle of proof **confirming or modifying the original requirements**.

To reiterate, designers need to know how users are likely to use a product in real-world tests with actual users, drawing on the results to confirm or modify the original design requirements. In a similar vein, the International Organization for Standardization's (ISO) *Human-Centred Design for Interactive Systems* (ISO 9241-210:2010) describes six key principles that will ensure a design is user centered. This involves (1) design clarity regarding the user, activities, and environments for use of the product; (2) user engagement throughout the process; (3) feedback drives the (4) sequential process that accounts for (5) a holistic user experience; and (6) which is inclusive of the rehabilitation team (International Organization for Standardization, 2010). The chief difference between these approaches and other product design philosophies is that UCD

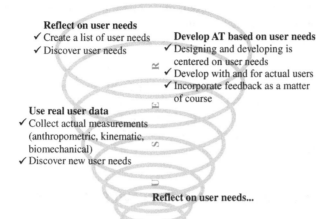

Reflect on user needs
- ✓ Create a list of user needs
- ✓ Discover user needs

Develop AT based on user needs
- ✓ Designing and developing is centered on user needs
- ✓ Develop with and for actual users
- ✓ Incorporate feedback as a matter of course

Use real user data
- ✓ Collect actual measurements (anthropometric, kinematic, biomechanical)
- ✓ Discover new user needs

Reflect on user needs...

FIGURE 10.1

User-centered design process.

tries to optimize the product around how users are able to use, prefer to use, or need to use the product, rather than forcing the users to change their behavior to accommodate the product.

Building on UCD principles, there has been a shift in perspective occurring at the collaborative edge of design and social science from a UCD process to that of participatory experiences. It is a shift in attitude from designing **for users** to one of designing **with users**. Participatory experience is more than just a methodology. It is based on the notion that everyone has something to contribute to the design process and that innovation and creativity are possible when users are provided with the appropriate tools and empowered to use them (Sanders, 2002).

PEOPLE WITH DISABILITIES AS INVENTORS AND COLLABORATORS

An often overlooked and discounted source of creative energy and ideas is the person with the disability or injury who will be the end-user of the technology solution. Assistive technology professionals, occupational therapists, nurses, and others who visit people with disabilities in their homes will attest to the large number of unique and creative solutions that people have come up with to perform their daily tasks. Sometimes these are designed by family members, friends, and neighbors but, quite often, these people are the fabricators of the designs proposed by the users. This shouldn't be surprising. The users have detailed and intimate knowledge of their own abilities, needs, and preferences. They often have a lot of time to visualize possible tools and solutions, proves websites, and network with other people with similar needs. In many cases, they have a solution worked out but need someone to help refine and implement it. Zebreda Dunham is an example of user-as-designer. She was the winner of the Rehabilitation Engineering and Assistive Technology Society of North America's 2015 Do-It-Yourself competition (Dunham, 2018).

THE CONSUMER-DRIVEN MODEL OF SERVICE DELIVERY

As we have seen, there are design models that strive to understand the needs of the user and others that incorporate the user in the design process. Assistive technology service delivery is inherently a consumer-centered process. The focus is not on advancement in the AT professional's career, to inflate egos, or to maximize billable hours. Many of us firmly believe that the best outcomes will arise out of a consumer-driven model of service delivery. Core components include

- The rehabilitation professional (whether engineer or disability-employment professional) acts as the **consultant** to the disabled consumer;
- The person with the disability is considered a **collaborative partner**, often leading the process;
- The consumer (i.e., the end-user) is the **expert** regarding their functional limitations and in activities in which they require help—they know best what works for them;
- The user can be a key **participant** in the creative problem-solving and design process;
- Assistive technologists and rehabilitation engineers should **avoid a prescriptive approach**, typically found in the medical field and not mandate solutions that the consumer has not tried or with which they may disagree.

OUR CHALLENGE TO DESIGNERS

- Embrace a consumer-driven approach to services, solutions, and design;
- Avoid designing *for* users in favor of designing *with* them; involve users early and often in the design process;
- Move from user-centered to user-driven and participatory design, seeking a collaborative relationship with the end user in the problem-solving and design process;
- Acknowledge people with disabilities as inventor, designers, and innovators;

- Work to promote these principles among emerging engineers and designers, especially among the volunteer and "maker" groups, and others without a background in AT and disability awareness.

The Bottom Line: the designer of custom accommodations for the workplace would do well to spend time discussing the project with the employee, getting an understanding of their priorities and preferences, and soliciting their ideas, suggestions, and feedback at various points in the design process.

THE ITERATIVE DESIGN PROCESS

It is important to understand and convey to others that design is an iterative process, meaning that it is rarely successfully completed in one attempt. Prototypes or initial design concepts are fabricated, tested out, and then tweaked, improved on, or abandoned and replaced, and tested again. At each step of the way, the user should be providing critical input as part of the UCD process discussed above. Obviously, the level of difficulty and complexity of the project will influence the extent of this effort.

DESIGNING FOR SHIFTING NEEDS

Another lesson learned the hard way by many designers is that the end-user's abilities, needs, and preferences can shift over time, and even from day to day. In some instances, the person's disability can progress, reducing their function. On the other hand, an injury can begin to heal, or medication can improve (or hinder) a person's functions. Some disabilities or illnesses, such as Multiple Sclerosis, chronic migraine headaches, repetitive strain injuries, or stress-related impairments like posttraumatic stress disorder can be quite variable. When possible, considerations should be made at the outset to mitigating triggers and irritants, such as light and noise levels, and providing flexible work tools, such as screen enlarging software for someone whose vision fluctuates. At other times, it is hard to anticipate changing needs or desires. For example, I once measured the preferred keyboard height for a person who wanted to stand at her computer and built a custom keyboard stand to accommodate her needs. When I delivered it, the employee decided that she would like it to be 4 in. higher. She said that her physical preferences can vary—something that she didn't tell me during our initial assessment. From that time on, I adopted the motto: "Make it adjustable, or make it again!"

UNIVERSAL DESIGN

Much has been written about Universal Design (UD) principles when developing products or spaces that can be usable by people of all sizes and abilities. Many of these can also apply to designing for the individual. For instance, the UD Principle of "Flexibility of Use" can be productively applied when designing for shifting needs as discussed above. Here is a list of the seven main principles developed by Ron Mace and others at the Center for Universal Design. Principles and guidelines are presented in their entirety from the Center for Universal Design:

The Principles of Universal Design

Principle 1: Equitable use
 The design is useful and marketable to people with diverse abilities.
Guidelines:
1a Provide the same means of use for all users: identical whenever possible; equivalent when not.
1b Avoid segregating or stigmatizing any users.
1c Provisions for privacy, security, and safety should be equally available to all users.
1d Make the design appealing to all users.
Principle 2: Flexibility in use
 The design accommodates a wide range of individual preferences and abilities.
Guidelines:
2a Provide choice in methods of use.
2b Accommodate right- or left-handed access and use.
2c Facilitate the user's accuracy and precision.
2d Provide adaptability to the user's pace.
Principle 3: Simple and intuitive use
 Use of the design is easy to understand, regardless of the user's experience, knowledge, language skills, or current concentration level.
Guidelines:
3a Eliminate unnecessary complexity.
3b Be consistent with user expectations and intuition.
3c Accommodate a wide range of literacy and language skills.
3d Arrange information consistent with its importance.
3e Provide effective prompting and feedback during and after task completion.
Principle 4: Perceptible information
 The design communicates necessary information effectively to the user, regardless of ambient conditions or the user's sensory abilities.
Guidelines:
4a Use different modes (pictorial, verbal, tactile) for redundant presentation of essential information.

4b Provide adequate contrast between essential information and its surroundings.

4c Maximize "legibility" of essential information.

4d Differentiate elements in ways that can be described (i.e., make it easy to give instructions or directions).

4e Provide compatibility with a variety of techniques or devices used by people with sensory limitations.

Principle 5: Tolerance for error

The design minimizes hazards and the adverse consequences of accidental or unintended actions.

Guidelines:

5a Arrange elements to minimize hazards and errors: most used elements, most accessible; hazardous elements eliminated, isolated, or shielded.

5b Provide warnings of hazards and errors.

5c Provide fail safe features.

5d Discourage unconscious action in tasks that require vigilance.

Principle 6: Low physical effort

The design can be used efficiently and comfortably and with a minimum of fatigue.

Guidelines:

6a Allow user to maintain a neutral body position.

6b Use reasonable operating forces.

6c Minimize repetitive actions.

6d Minimize sustained physical effort.

Principle 7: Size and space for approach and use

Appropriate size and space is provided for approach, reach, manipulation, and use regardless of user's body size, posture, or mobility.

Guidelines:

7a Provide a clear line of sight to important elements for any seated or standing user.

7b Make reach to all components comfortable for any seated or standing user.

7c Accommodate variations in hand and grip size.

7d Provide adequate space for the use of assistive devices or personal assistance.

The Principles of Universal Design used with Permission: North Carolina State University, The Center for Universal Design (1997).

The Principles of Universal Design were conceived and developed by The Center for Universal Design at North Carolina State University. Use or application of the principles in any form by an individual or organization is separate and distinct from the principles and does not constitute or imply acceptance or endorsement by The Center for Universal Design of the use or application.

INCLUSIVE DESIGN FOR SOFTWARE AND WEBSITES

There may be situations where a custom software application or a smartphone app need to be developed as part of an accommodation, or when several competing products may need to be evaluated. At other times, the AT service provider may need to create instructions or training materials. Designing for people with disabilities with a particular focus on websites and software applications has been promoted as "Inclusive Design." Resources related to this can be found at

- Microsoft's inclusive design principles: https://www.microsoft.com/en-us/design/inclusive
- Inclusive design principles (similar to UD): http://inclusivedesignprinciples.org/

The following tips can be helpful when organizing training materials or designing equipment interfaces.

Inclusive Design Principles

- Provide an equivalent user experience: Make sure all users are able to effectively perform their work tasks without negatively impacting the quality of their work.
  - Context matters: A core consideration here is an effective user experience no matter where they are engaged in work tasks.
  - Eliminate variability: Standardize and do not deviate from your approach.
  - Encourage user preferences: Avoid dictating user access—user preferences should be maximized.
  - Be flexible: Consider multiple means of task engagement—allow users to choose how they engage tasks to increase flexibility and efficacy.
  - Organize around purpose: Reduce information overload and attention fatigue by organizing information around the purpose for its presentation.
  - Focus value-added features: Further improve flexibility and choice by considering multiple means of user interaction—this drive user satisfaction (Swan, Pouncey, Pickering, & Watson, 2018).

KEEPING IT SIMPLE

Although this concept seems obvious, too often designers are drawn to solutions that are more challenging or interesting. The simpler and more basic the solution, the lower its cost and the shorter its implementation time. Additionally, repairs and redesign are easier. Along these lines, it is interesting to note how often groups of volunteer engineers, student design teams, and handymen and women "reinvent the wheel" by designing and building items that are already available in the marketplace. While sometimes this is an effort to come up with lower cost solutions, it is no doubt often driven by a desire to invent and make something rather than to just raise the money to buy it. Engineers and AT specialists who work in the field are not immune to the lure of an opportunity to express their creative energies in this way. They just need to keep in mind that someone has to pay for the end product, while someone else is waiting to get it in a timely manner, so they can do their job. TIP: When you hear a designer talk about the KISS principle, they are referring to the concept, "Keep It Simple, Stupid."

CASE STUDY: CASE EXAMPLES OF SIMPLE SOLUTIONS

An office worker with reduced reach and hand use needed to read printed manuals on occasion. Her power wheelchair kept her from getting closer to her desk to reach the manuals, and she tried balancing them on her lap, with limited success. She had another desk for computer access, and there were concerns about the cost of getting a second new desk just for this occasional reading task. I noticed that she had a large calendar on her desk and suggested that she try sliding it out onto her chair's armrests to provide a support for a bookstand. This was promising but not strong enough, so I came back with a thin wooden board to place under the calendar. Problem solved, with minimal time and expense.

In another example, a person working for a social service agency spent part of her time at a senior center where she helped people fill out necessary forms. She worked at an old desk in a cramped space and others used the desk when she was not there. She had back problems that were exacerbated by her twisting constantly in her straight-back chair to look at clients who were seated at the end of the desk. The first part of the solution was helping her obtain a more suitable chair. Additionally, a fabricated wooden surface (i.e., a drawer insert) that she could place on the pull-out drawer on the side of her desk created an L-shaped desk configuration that allowed her to position herself so she could face clients while assisting them. When she left for the day, she simply stowed the insert underneath the desk and closed the drawer.

FABRICATING TECHNOLOGY SOLUTIONS

There are some important distinctions between general product design and designing for a job accommodation. Consumer products are manufactured and sold with a large user base in mind. Even in a workplace, general changes to an office layout, manufacturing process, or assembly line are made (with more or less success) to suit the majority of workers. On the other hand, as we have already made clear, a job accommodation entails designing for the individual. Therefore, each one can be unique in their own ways. What does this mean in practice for the fabricator? *Note:* In this discussion, we will assume that the designer and fabricator are the same person, though some larger programs employ AT specialists or rehabilitation engineers who come up with detailed specifications and leave it to skilled technicians or fabricators to implement them.

THE COST OF FABRICATION

A prime consideration when beginning a fabrication project is cost—both time and materials. Employers may or may not have a lot of resources and may be reluctant to provide expensive accommodations for a new employee who may or may not work out. (The attitude and available funding can be very different if a skilled and seasoned employee needs to return to work after a disabling injury or an effort needs to be made to help somebody with a developing disability maintain her job.) Often an insurance company may be involved. Third-party funders such as vocational rehabilitation agencies have limited funds for services that must be spread over many consumers. Whatever the situation, cost is always part of the discussion.

Cost in time. It is always difficult to project how long a design or fabrication project will take. What might seem at the start as a "minor" element, might take hours to resolve because of the limitations introduced by other design parameters. Quick prototyping using simple materials is often a good way to begin. This can range from what my associate calls "Basic CAD: Cardboard Aided Design" to mockups using sheet plastic, wood, or 3D printing that can easily be modified or tossed,

Time costs have an additional component besides labor charges for the design and fabrication. This includes the lost time in productivity to the employer who wants an employee to return to their job or to get a new hire fully online and up to speed. Of course, from the employee's end, he or she wants to get their full paycheck as soon as possible. In these situations, the clock starts ticking, and the pressure is on to deliver something as soon as possible.

Cost in materials. The choice of materials is a multifaceted question. Often the design specifications, environment, available space, and required tolerances will dictate the materials utilized in the fabrication. At other times, the same general solution can be achieved using different materials and techniques. In these situations, the choice of materials will often depend on the fabricator's skills, comfort level, and available tools. The cautionary note here is not to "over engineer" a solution. Just because one has a milling machine, does not mean that the part needs to be machined out of a solid block of aluminum when one made from plastic glued together will do the trick. When welding or otherwise fabricating a bracket out of steel, one has to factor in the time and cost of cleaning and coating the steel to keep it from rusting. While 3D printing is becoming increasingly popular, at least among those with limited shop skills, factors to consider include the fabricator's abilities and skills with design software, delays introduced by the slow printing time, and limitations of the materials available with more affordable printers. (Admittedly, the attraction of using this and other new technologies may be hard to resist by the fabricator.) On the other hand, the quicker solution must be durable enough to withstand potential stresses and wear over time, so it does not have to be repaired or replaced too soon.

STAGING THE SOLUTION

Most complicated design and fabrication projects are inherently iterative. That means that it involves a bit of trial and error and requires testing prototypes both on the bench and with the end-user, refining the design until a final product is delivered. Hopefully, the design specifications were clear enough, the information gathering and input from all the interested parties was sufficient and accurate, and the skills of the fabricator were appropriate, so that the process could be a reasonably efficient one, with few trips back to the shop and even fewer trips back to the drawing board. As already noted, this can be made easier by initially fabricating mockups that are easy to adjust and change on site or rough prototypes that require less time to produce.

While most people would agree that the "look" of the final device or accommodation is important for user acceptance and can minimize the extent to which it is a focal point of attention, this is a highly subjective component. Some users care a lot about cosmetic features such as color, while others just care that the device is functional. At the same time, it should be clearly conveyed during the prototyping phase that what might appear to be rough-cut will be more polished in its final iteration.

LOCATING RESOURCES

Not everybody has ready access to an AT specialist, rehabilitation engineer, or other related professional who can do fabrication. This is especially true in more rural environments or locations farther from the larger cities. In these cases, there are various possible sources, including the employer's own maintenance and repair facilities, local machine or welding shops, carpenters or cabinetmakers, and general repair and handyman services. Staff at better hardware stores might have suggestions. Local service organizations can be contacted to see if any of their members are available for hire or interested in donating their time. (One possible drawback with seeking assistance from service organizations, associations of retired engineers, or engineering or design classes at local colleges and universities is that the need for a solution in a work setting may be urgent and doesn't always mesh with the time and resources available to volunteers or students.) In instances where the fabricator may not have been involved in the preliminary solution or design process, it is even more important that the initial work be done with adequate consideration and attention to detail.

SUMMARY

During the assessment and problem-solving process, information is gathered and solution criteria are established that inform the selection of assistive technology devices and, when needed, the modification of existing items or the design and fabrication of new ones. To do this custom work successfully, the designer/fabri-cator—or the AT service provider directing them—will benefit from following well-established design principles such as User-Centered Design and Universal Design. Respecting the hierarchy of possible solutions and acknowledging the real-world demands of cost and time constraints will help keep things on track. Involving the end user to the extent possible and embracing a consumer-driven model of AT service deliver will not only empower the consumer, but will also enhance the appropriateness, quality, and usability of the implemented solution.

REFERENCES

Czaja, S. J., & Nair, S. N. (2012). Human factors engineering and systems design. In G. Salvendy (Ed.), *Handbook of human factors and ergonomics* (4th Ed., pp. 38–56). Hoboken, NJ: John Wiley & Sons, Inc.

Dunham, Z. (2018) Zebreda makes it work. Retrieved from http://www.zebredamakesit-work.com.

International Organization for Standardization. (2010). Ergonomics of human-system inter-action—Part 210: Human-centred design for interactive systems. Retrieved from https://www.iso.org/obp/ui/#iso:std:iso:9241:-210:ed-1:v1:en.

North Carolina State University, The Center for Universal Design. (1997). The principles of universal design. Retrieved from https://projects.ncsu.edu/ncsu/design/cud/about_ud/udprinciples.htm.

Perkins, D. N. (1986). *Knowledge as design*. Hillsdale, NJ: Lawrence Erlbaum Associates, Publishers.

Sanders, B. N. (2002). From user-centered to participatory design approaches. In J. Frascara (Ed.), *Design and the social sciences* (pp. 1–8). New York, NY: Taylor & Francis Books Limited.

Schwartz, B. (2015). *Why we work*. New York, NY: Simon & Schuster.

Swan, H., Pouncey, I., Pickering, H. & Watson, L. (2018). Inclusive design principles. Retrieved from: http://inclusivedesignprinciples.org/.

Usability.gov. (2018). What and why of usability: User-centered design basics. Retrieved from https://www.usability.gov/what-and-why/user-centered-design.html.

Implementation, training, support, and maintenance

11

Ray Grott[1] and Anthony Shay[2]

[1]*Rehabilitation Engineering and Assistive Technology (RET) Project, San Francisco State University (SFSU), San Francisco, CA, United States* [2]*Capacity Building Specialist, Assistive Technologist, and Rehabilitation Specialist, University of Wisconsin-Stout Vocational Rehabilitation Institute (SVRI), Menomonie, WI, United States*

INTRODUCTION

The provision of assistive technology (AT) is more than just ordering the items identified in the assessment process and delivering them to a consumer. The AT needs to be inspected upon receipt and may be trialed/tested by the AT professional, delivered to the consumer, assembled, installed, and configured or adjusted. The implementation may take place in any number of locations as deemed appropriate by the consumer, the referral/funding source, AT Professional, and other rehabilitation team members. Training may follow implementation as a stand-alone service, or it may be provided as part of the implementation process. Training may occur following a significant lag in time after an assessment. It may be "once-and-done" or may be provided over several meetings—in person or remotely. Keys to effective training include flexibility, patience, and developing and maintaining a good working relationship with the user of the AT.

ASSISTIVE TECHNOLOGY IMPLEMENTATION: AN OVERVIEW

Fig. 11.1 illustrates that people enter the assistive technology service delivery process at different points—broadly speaking—they may begin with an assessment, the implementation phase, or with training. Regardless of the starting point, cases are typically opened with a referral and an intake. We make the assumption here that implementation is following an assessment, and that the recommendations in the report have been approved by the referral/funding source, and the implementation funding has been secured.

When the implementation phase begins, the questions received from referral/ funding source during the assessment process are reviewed, and any additional questions the AT professional has are sent to the referral/funding source for

Assistive Technology Service Delivery. DOI: https://doi.org/10.1016/B978-0-12-812979-1.00011-4

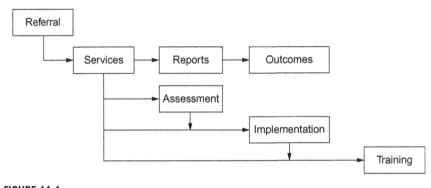

FIGURE 11.1

Training as a stand-alone service or as a component of the AT implementation process.

clarification before moving forward with the case. The consumer is also contacted to gather any additional information needed to move forward. An implementation plan is completed and sent to the consumer. With their approval, AT procurement and development can begin. An implementation plan typically consists of a brief letter outlining when the AT assessment was completed. The letter lists the date of the approval received from the referral/funding source, the authorizing agent, and approval for the implementation to be scheduled and conducted. The implementation plan typically breaks down projections for completion dates for AT ordering; set-up and configuration; and delivery, installation, and training.

The procurement and development step in the process encompasses the purchase of any off-the-shelf commercially available products, parts, or materials necessary to meet consumer needs as well as customization or fabrication of an AT solution. Then, the off-the-shelf items and/or the customized/fabricated AT is received by the AT professional. The consumer may have been contacted and/or met with several times in the preceding period to review specifications, trial/demonstrate, and/or adjust any customized/fabricated AT to ensure the appropriate fit with consumer needs. The consumer is then contacted to schedule an on-site implementation date and the referral/ funding source is notified. Some agencies may order commercial items and have them sent directly to the consumer. However, problems can occur when boxes are opened and devices or parts are lost or broken. In addition, order or shipping errors and damage in transit may not be discovered in a timely manner when the items are shipped to the consumer. Having items shipped to the AT service provider's address so they can be inspected, organized, and configured properly may be advisable.

Once received, the AT items are loaded and taken to the consumer's preferred location (e.g., home, work site, referral/funding source's location, school library, etc.) for deployment. Initial training occurs at this meeting, which may be sufficient for the consumer. If additional training is required, training funding is secured from the referral/funding source, and additional training dates are scheduled with the consumer.

The consumer is contacted shortly after the last on-site or remote meeting to allow for an opportunity to troubleshoot any problems and to notify them of any impending changes to AT service case status. Persistent problems would result in additional meetings and, if necessary, additional training. A final report is prepared and sent to the referral/funding source. Depending on the complexity of the case, the referral/funding source may want a final meeting with the consumer, the AT professional, and/or the rehabilitation team. The AT file is then organized, and purchases reconciled (purchase receipts for items versus provided items list). If the case is closed, the files may then go through a quality assurance process and then into storage.

TECHNOLOGY PROCUREMENT AND DEVELOPMENT: TOWARD PROPER IMPLEMENTATION

An ideal product or an ingenious design idea will not be successful unless it is properly implemented. There are a number of elements to consider for effective implementation.

To move forward with a technology solution that requires components that are not already available, somebody has to pay for the necessary items. While this may seem obvious, it often causes the project to be delayed for a significant amount of time. Additionally, the provision of rehabilitation engineering, assistive technology, and fabrication services are typically provided by individuals or businesses on a fee-for-service basis. On the employer's end, the purchasing process and authorization process can be straightforward and efficient, or it can require multiple levels of approvals and bureaucratic hoops. Some public agencies are required to utilize preferred vendors and it can take extra time and effort to get approval to work with others. Often contracts need to be written and signed, with supporting documents such as proof of liability insurance attached. If a vocational rehabilitation or other agency is involved, they will have their own rules and regulations for purchasing equipment and authorizing services.

There are various things that a service provider can do to facilitate this process. To begin with, they should become clear early on about who is expected to cover the costs of a recommended accommodation. While in the United States it is technically the responsibility of the employer to pay for the accommodations of an employee (i.e., consumer) with the disability, an expensive solution might be viewed as an "undue hardship," as discussed in the previous section, and rejected or denied. In addition, these rules do not apply to companies with fewer than 15 employees. This is another argument for keeping the costs down, as well as a reminder that the employees in the United States can seek support from their state's vocational rehabilitation office before they seek employment or if they are trying to retain their job. If the person is a military veteran, they may also qualify for employment supports through the relevant Veterans Administration program.

It often falls to the AT service provider to make sure that all parties are talking to each other, since the consumer is usually unclear about the procedures and how they can get their needs met. The service provider is frequently asked by the consumer to provide guidance on policies and procedures as well as their legal rights. They should therefore be prepared to direct the employee to the company's human resources office or to government and nonprofit agencies with staff who are knowledgeable on accommodation issues. In addition, the service provider may need to encourage the consumer to be a stronger self-advocate and suggest that they discuss their concerns and fears with rehabilitation team members.

When discussing possible solutions with the employer or other funding sources, it is always good to speak openly about possible total costs to avoid last-minute surprises. After an assessment, the AT service provider should provide a detailed, written report that outlines the need and justification for any recommended technology. To facilitate the approval and procurement process, the report should include specific product names and manufacturer part numbers, along with estimated prices (if requested). It should also identify the estimated costs for designing and fabricating any custom solutions and installing and training in the use of the technologies. The evaluator should also be prepared to explain in more detail why a recommended item is a necessary and essential component of a needed accommodation and to rank the items in order of priority.

PUTTING IT ALL TOGETHER

Once the items have been purchased and delivered and any custom items fabricated, it becomes the time to start using them. However, there is more to the implementation process than simply providing assistive technologies and closing a case. Rather, the purchase and receipt of AT items is only beginning. Following delivery to the consumer, the technology needs to be assembled, installed, configured, or adjusted, depending on whether it is software or some type of hardware. Too little time allowed for adjusting and dialing-in the system for the user can lead to frustration and an attitude that it won't work. As in every aspect of the service delivery process, good communication is important—you want to know as soon as possible if something is not meeting expectations, rather than finding out a month or two later that it is not being used, and the consumer (employee) didn't report it because they did no longer want to disappoint anyone, thought it was their only shot at getting something, was unsure who to contact, etc. (We'll say more about the discontinuance of technology use later in this section.)

ASSISTIVE TECHNOLOGY TRAINING

It is likely that the AT service provider will be very familiar with the off-the-shelf devices and software they recommended and will know all the details of the custom ones they designed, programmed, or fabricated. Not so for the person who will be using it. Adequate training therefore becomes a critical component for successful implementation. This sounds good on paper, but it can easily be a

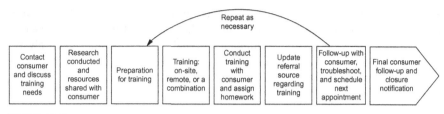

FIGURE 11.2

Training detail in the AT implementation process.

weak spot when the employer is eager to get a full day's productivity from the employee, the employee may feel they are too busy to spend the time required to learn new things, and the AT specialist is off working on new projects. In addition to face-to-face training, presented in a learning style and pace that the user can handle, written instructions and tip sheets should be provided as a reference if all of the new information wasn't retained. Training may be provided as a standalone service or as a component of the implementation process. A significant amount of time may have elapsed between an AT assessment and a referral/funding source request for training. In these instances, training would occur barring any intervening factors necessitating the need for a new assessment. There may also be times when the employee, a disability-employment professional, family, or the employer have already purchased an AT device or system, and training is determined to be all that is necessary. As training needs are determined to have changed, additional research and consumer contact may be necessary. We go into more detail regarding training and possible guidelines later in the chapter (Fig. 11.2).

EVALUATION AND REVISION

As we mentioned earlier, the design of a solution—ether using custom or off-the-shelf components—is an iterative process. We have to see how close the combination of items comes to working as desired. If the modification or custom device is complicated or if the design parameters are a bit unclear, it is usually good to produce a rough prototype for the employee to test before putting the time and effort into a finished product. After the user has had a chance to work with it, and the evaluator has been able to observe it being used, adjustments, minor tweaks, or significant design changes may be needed. Sometimes a rethinking of the design or solution is required because a key consideration was overlooked, but hopefully that will be rare if the problem analysis was done properly. It's also possible that, once the device is deployed, one of the parties might visualize a different and more efficient way to design the device or approach the solution. If it is significantly better, that insight should be welcomed; bolstering the argument for starting with a prototype when appropriate. It would be ideal to allow the user plenty of time to get used to and evaluate the functionality of the product or design before the case is closed, although cost and business realities don't always allow that.

FOLLOW-UP SERVICES

Just as we don't buy a car and never do anything but put gas in it, a technology solution shouldn't be implemented without a plan in place for providing necessary maintenance, troubleshooting, repairs, and additional or remedial training. Often this can be accomplished by having the provider return to make sure everything is adjusted and working properly or, more realistically, being able to call on them when something goes wrong. If the service provider is hours away and local resources are not identified, a small problem that can be readily fixed by someone with the right know-how might end up sidelining the whole system. The availability of maintenance and repair services should therefore be a consideration when selecting the best solution to a problem.

ASSISTIVE TECHNOLOGY TRAINING WITHIN EMPLOYMENT SERVICE DELIVERY FOR PEOPLE WITH DISABILITIES

In disability-employment systems training tends to bifurcate along two lines: train-then-place and place-then-train. When we look for work, we either provide training and then obtain work or we place an individual and train them on-the-job, respectively. The train-then-place approach relies on a combination of hands-on and more formal instruction (e.g., classroom). These activities take place prior to job development and are based on goodness-of-fit between an individual and their Great 8. Because of the formal training component, consumers with developmental and cognitive impairments may have difficulty benefitting from this type of program due to difficulties with knowledge transfer (Fadely, 1987). A train-then-place approach does not exclude training on a job site. However, emphasis is placed on the provision of necessary training early in the process with follow-up training provided on the work site once a job is obtained. Immersion in the work context and task engagement activity can be gradual or with full immersion quickly based on an individual's needs. As mentioned earlier, the severity and type of impairment can impact how this occurs. The duration of job development activities is typically uncertain so planning for training can be difficult. Train-then-place advocates may emphasize the need to provide comprehensive services in as controlled a manner as possible arguing that full immediate immersion in the work context with its commensurate demands, and stressors can exacerbate or cause disabling conditions that might impair functioning and jeopardize the job (Corrigan & McCracken, 2005).

Place-then-train advocates would argue against this assertion. Train-then-place is generally viewed as coming from a medical model of job placement whereby consumers have barriers or impairments to employment that must be addressed (or fixed) before attempting to enter the job market. Inadequately addressing these needs leads to either a problematic job search and/or eventual job separation. Corrigan and McCracken state that the medical model is built along a "continuum of care" which emphasizes a supervision and support structure at each step along the way on the continuum. It might be argued that people with disabilities moving through a sequential process like this will experience a disjointedness derived from this stepwise movement. This "chronic dislocation" tends to impede stabilization. Advocates of the place-then-train approach argue that the phase-based process of train-then-place approaches is difficult for anyone and that greater immersion more quickly is preferential to a slow and deliberate methodical process (Corrigan & McCracken, 2005, p. 32).

With the advent of the rapid engagement approach to finding work for people with disabilities, it is likely that assistive technology providers will find themselves more often than not meeting with people within employment contexts. Corrigan and McCracken suggest that the place-then-train approach to training is inherently empowering. Offering people the capacity for on-the-job training and work adjustment facilitates gains in independence, self-reliance, and vocational competence. This approach works well for individuals with mental illness. Corrigan and McCracken (2005, p. 334) state that "relatively rapid placement in vocational ... situations is essential for helping persons achieve their goals. Only then do the elements of training programs have any meaning." Corrigan and McCracken (2005, p. 32) find that place-then-train models can offer greater opportunities for the development of "coping and adaptation strategies" in work contexts and address issues related to others in the work environment including those of prejudice and discrimination, all of which can give rise to psychological distress and adjustment problems. Place-then-train models can offer people with disabilities quicker access to the world of work, a jump start on integration, all while capitalizing on the initial enthusiasm around entry or reentry into employment (Bond, Drake, Mueser, & Becker, 1997). Real-world access to a job site is a critical aspect of providing assistive technology services to people who need them to effectively perform work tasks making a place-then-train method for AT implementation and training a practical approach.

PERSON-CENTERED TRAINING

Whenever we work with an individual who has a disability, regardless of the condition or its severity, we must hold fast to the idea that they have the capacity for growth beyond the initial point at which we find them. Just as the assistive technology service delivery process up through the implementation step has been person-centered so too are the implementation and training stages of service delivery (i.e., every step in the process). Whenever assistive technology service provision is not reflective of individual needs, service providers sacrifice effective assistive technology use, job meaning, job satisfaction, and programmatic outcomes.

The U.S. Department of Education offered three useful points related to person-centered training strategies in the *Final Report of the National Mathematics Advisory Panel*:

- Training should include both explanation and demonstration including consumer verbal problem solving;
- Task elements being trained should be deliberately ordered toward identification of the primary elements of the task;
- Consumer progress should be monitored and provided regular and explicit feedback related to successful completion of the task elements being trained (Greeno & Collins, 2008).

Of course, the level of assistance a consumer will require is variable and will change over time. As tasks are learned, training is necessarily faded. Fading is relative to functional limitations and may not be faded entirely (as in supported employment and use of a job coach to monitor and train long-term). Self-doubt, mistakes, and complacency can have an impact on efficiency (MacDuff, Krantz, & McClannahan, 2001)—this has been referred to as psychic entropy (Csikszentmihalyi, 1990) and motivational drift (Hebb, 1949, p. 209). Training strategies may include one or more of the below methods (list is not exhaustive) (Table 11.1):

Table 11.1 Employment-Related Assistive Technology Training Strategies

1. Delayed prompting	11. Direct Verbal Prompting/Direct Auditory Cueing
2. Foreshadowing	12. Indirect Verbal Prompting/Indirect Auditory Cueing
3. Using Gestures and Nonverbal Behavior	13. Full Physical Assistance
4. Ignoring (avoiding overt attention)	14. Partial Physical Assistance and Physical Priming
5. Instruction (written or verbal)	15. Incremental Prompting or Graduated Guidance
6. Role Playing and Simulation	16. Modeling and Demonstration
7. Using Naturally Occurring Cueing	17. Monitoring (overt or covert)
8. Positional (corrective) Prompting	18. Positive Reinforcement (affirmation)
9. Using Repetition	19. Silence (promoting self-exploration and introspection)
10. Visual Prompting	

Determining an effective training strategy requires flexibility, patience, and good working relationship with the consumer and the referral/funding source. Effective training, according to psychologist Bruner (1996, p. 119), "requires a task that has a beginning and some terminus". This seems obvious, however, explicitly framing tasks for people—making explicit when they begin, the steps in-between, and when they end—helps build awareness, meaning, and job satisfaction. Having some guidelines for training helps structure the process for those we work with in AT service delivery.

TAKING A CLOSER LOOK AT JOB TASKS

Learning is oriented around building knowledge and understanding. As we mature, three conditions facilitate learning. We learn to

1. Structure and organize concepts, separating them from the actions they imply,
2. Separate affective attachment from concepts, and
3. Separate the sense of achievement from any consequences—focus on outcomes as information only.

This "denatured" approach to instruction provides a supportive and encouraging atmosphere conducive to learning (Bruner, 1966, p. 134). It is also a thumbnail sketch of a task analysis process toward task acquisition.

Developing effective training strategies begins with understanding the tasks a consumer is expected to perform in the context in which they take place. A task analysis is the formal process by which this occurs. A task analysis facilitates task acquisition through the descriptive analysis of the task engagement activity, delineation of expected outcomes (i.e., goals and objectives), job coaching strategies (i.e., teaching and training approaches, consumer characteristics), and environmental considerations. It may be helpful to begin with sketch of the work environment to map out the physical layout of the task engagement activity. A task analysis involves the breaking down of a task into its constituent parts. This allows for each component to be understood as a unit and as a part of the entire task. A task flow analysis may also be beneficial as a means to situate the task being analyzed within the work flow processes in which it is located (Rother & Shook, 2003). Care should be taken to ensure the sensory, physical, cognitive, language, and contextual needs of the consumer are met as they relate to each task they must perform.

A visual model is also helpful in breaking down tasks. This type of modeling can be developed in many ways depending on the needs of the consumer. Pictures or icons may be used instead of or in conjunction with text. Color or shading may be used to differentiate components. The trainer and consumer can have corresponding versions of the task breakdown. The trainer's copy can be built-out in greater detail to facilitate teaching or guiding the consumer. The consumer may benefit from greater detail. The detail in consumer's version of the breakdown should be reflective of functional skill levels.

The depth and detail may be extensive. For example, the "Get coffee from the pantry" element may have five steps under it reflecting that the coffee is in the pantry, second shelf from the top, the coffee is behind the hot chocolate, kept in four different containers based on flavor, select the flavor of the day as marked on the container, and take the coffee to the coffee pot. As consumers learn these steps, the model can be modified to present only the elements necessary to accomplish the steps. A consumer may only need the first row to complete the task. Detail and training is reduced as the task is learned with appropriate accommodations in place. This builds independence on the job rather than dependence on the trainer, coworkers, or the instructional tools.

Fig. 11.3 illustrates a basic task map for the "making coffee" task, and how this fits into a work flow process. The map could break down only the making coffee task to illustrate the task elements for the **process time** (PT), **completeness and accuracy** (C/A) with which the task was completed, and the **wait time** between each element. This helps the trainer determine where problems may arise. Should the trainer have an opportunity to monitor the processes on either side of the consumer, issues outside of the consumer's control may be identified forestalling any objections coworkers or management may have regarding consumer performance.

In this example, the consumer's (employee/barista) task is making coffee. She must wait 15 minutes for the previous task to be completed; it takes 15 minutes for the donuts to be displayed. This task is completed with 92% accuracy. This is a relatively high percentage of accuracy. Accuracy of less than 100% may be the result of a donut being dropped, fingerprints in the frosting, or putting out day-old instead of fresh donuts. Guests may complain to the barista that the donuts do not meet their quality standards or expectations. The result being, at a minimum, the PT for making coffee increases due to interruptions. There is a 2-minute wait following the donuts being displayed before the coffee can be made. It takes our barista 15 minutes to make the coffee that another coworker then serves. Accuracy and PT in the previous and following tasks, being performed by one or more coworkers, may impact the consumer's (employee's) job or the perception of how efficiently they are performing their duties. In our example, displaying donuts is completed at 92% C/A, and refilling patron mugs is completed with 85% C/A. Taken together with making coffee completed by our barista with 95% C/A, the average for those tasks considered together is 91% C/A. An employer expecting no less than 95% C/A across all three tasks may view the coffee maker as meeting expectations. We can demonstrate otherwise. Looking at the work process flow can facilitate a better understanding of how each task impacts our consumer and their task(s) and that of the coworkers around them (see Fig. 11.4). Task mapping, flow, and graphing would be completed for each specific task/element.

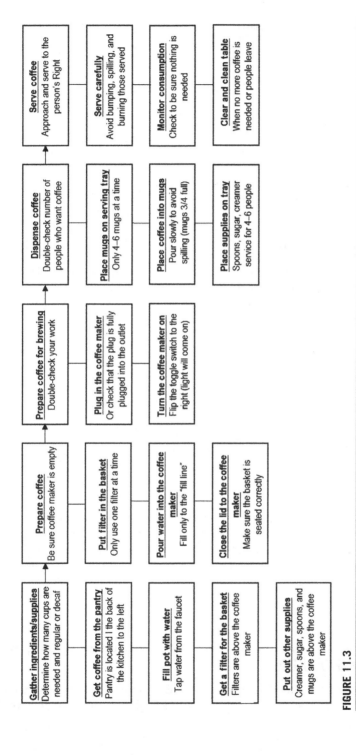

Gather ingredients/supplies
Determine how many cups are needed and regular or decaf

Get coffee from the pantry
Pantry is located l the back of the kitchen to the left

Fill pot with water
Tap water from the faucet

Get a filter for the basket
Filters are above the coffee maker

Put out other supplies
Creamer, sugar, spoons, and mugs are above the coffee maker

Prepare coffee
Be sure coffee maker is empty

Put filter in the basket
Only use one filter at a time

Pour water into the coffee maker
Fill only to the "fill line"

Close the lid to the coffee maker
Make sure the basket is seated correctly

Prepare coffee for brewing
Double-check your work

Plug in the coffee maker
Or check that the plug is fully plugged into the outlet

Turn the coffee maker on
Flip the toggle switch to the right (light will come on)

Dispense coffee
Double-check number of people who want coffee

Place mugs on serving tray
Only 4–6 mugs at a time

Place coffee into mugs
Pour slowly to avoid spilling (mugs 3/4 full)

Place supplies on tray
Spoons, sugar, creamer service for 4–6 people

Serve coffee
Approach and serve to the person's Right

Serve carefully
Avoid bumping, spilling, and burning those served

Monitor consumption
Check to be sure nothing is needed

Clear and clean table
When no more coffee is needed or people leave

FIGURE 11.3

Task mapping diagram for making coffee.

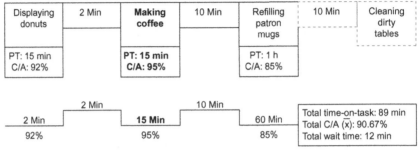

FIGURE 11.4

Task flow mapping diagram for making coffee.

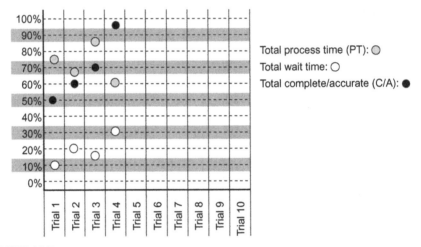

FIGURE 11.5

Graphing progress from task mapping.

Following task mapping the observed results can be recorded to better measure progress and spot patterns through successive trials and work experiences as the consumer learns the task and as the trainer (or supervisor, job coach, or evaluator) monitors and assesses task activity. Fig. 11.5 illustrates task graphing for progress.

TRAINING: GUIDELINES FOR PLANNING, ORGANIZING, AND MEANING-MAKING

The following three training processes primarily utilize a place then train modality, though not necessarily exclusively. Each process consists of several key points and objectives. Together these represent loose guidelines for approaching task training. First, we work to understand our work tasks and those it impacts— we appreciate as best we can the lay of the land. We seek also to break the task down into smaller parts. Looking at the big picture and then focusing in on the details provides us a comprehensive working knowledge of the task expectations and purpose; how the work we do fits into the larger scheme of things. Next, we take stock of our personal and learning resources, and what we need to do toward being an effective learner. Finally, we immerse ourselves in the task. We learn how to use any accommodations effectively and gradually achieve full immersion and task engagement in the work context.

Process one: Mapping it out.
1. Know your target task or activity.
   a. Frame it in the form of a question or statement. You must know what you want to achieve to know when you have achieved it (e.g., I want to learn how to/How do I make and serve coffee in a café setting).
   b. What is the task or activity which is the focus of the training?
   c. Consider reviewing the functional elements of employment (discussed earlier in the chapter) in their entirety before proceeding with task acquisition activity (providing a big picture view before detail work begins).
   d. Write down or map out the objective (observable) tasks and task elements (as opposed to the subjective elements such as assumptions we make or how we feel) (see Fig. 17.3).
2. Deconstruct the task or activity.
   a. Understand that a job consists of many tasks. A task consists of many parts. To learn a task involves learning the parts.
   b. Break the task down into its constituent parts (e.g., task elements).
   c. Determine which parts are most salient to your needs. Start with these first.
   d. "Denature" the task elements by removing the implied action, affective component, and notion of consequences from learning the elements of the task (removes the stress, anxiety, and reward/punishment aspects of actual work task engagement until after the task elements/tasks are learned).
   e. Consider learning the parts in a different order than usual. The parts you do learn can be built upon.
   f. Consider separating task elements into simple and complicated (or graduated) difficulty. Begin with the simpler task elements first. Build on successes.
3. Identify the core features of the task or activity.
   a. Look for the most frequently used features or qualities—task element repetition.
   b. Look for core features or task elements within the task.

    c. Look for patterns, arrangements, and rhythms.

    d. Map the work environment to better understand the physical layout and spatial relationship between features in the task engagement context(s).

Process two: Resource mobilization

4. Mobilize personal resources.

    a. Take a personal inventory of your Great 8 affirming and reinforcing personal agency toward the task training goals.

    b. Reach out within your Circle of Resources to ensure basic needs are met.

    c. Resolve to put in the time and effort to learn the task or activity. Persevere.

5. Utilize learning resources.

    a. Obtain training content in more than one format. Find sources of information in as many mediums (visual, auditory, kinesthetic) as you need or are available.

    b. Develop written instructions or tip sheets to facilitate instruction and learning.

    c. A key to memory retention involves using all the senses as a means to encode a memory. Think it, feel it, smell it, hear it, taste it, visualize it. Experience the parts of the task you are trying to learn. The more sensory modes you use to encode a memory the more likely it is to be retained.

    d. Utilize different learning styles (visual, auditory, kinesthetic) to optimize and reinforce learning.

    e. Do not allow yourself to be overwhelmed by the sheer amount of information that may be present. Pace yourself.

    f. Begin with a limited amount of information. Chunk it out.

    g. Balance functional capacity load-bearing with extant skill levels (skill levels that still exist following the onset of disability)—the balance will change over time with practice and accommodation.

6. Establish an environment conducive to learning.

    a. Work to eliminate distractions and items competing for your attention.

    b. Consider off-site training of task elements that can be done prior to job placement and which will not negatively impact on-site task engagement (Wehman, Inge, Revell, & Brooke, 2007).

    c. Be aware of environmental issues (e.g., lighting, noise, clutter, accessibility, tool usability).

    d. Obtain the necessary assistive technology (or other accommodations) toward overcoming functional limitations.

    e. Address any other barriers inherent in the training and task engagement processes.

    f. Consider universal design principles for products, the built environment, instruction/learning, etc.

Process Three: Task Engagement

7. Establish goals and objectives (outcomes and the changing intermediate goals that are modified through feedback).
   a. Intermediate goal setting and intrinsic motivation reinforce task engagement behavior.
   b. Be aware of the affective determinants of behavior. Be aware of how you feel and the circumstances around this feeling—the quality of the positive experience. This can reinforce task engagement and motivation.
8. Establish a baseline. Although this may not be a necessary part of the learning process, it can be difficult to see what progress has been made (or what ground has been lost) without knowing where you started.
   a. Practice, practice, practice. Repetition reinforces learning.
   b. Practice also involves visualizing one performing the task. This also counts as physically practicing the task (though not replacing physical task engagement).
   c. The focus is not on being perfect; rather, it's on learning the basics. You can build on and refine them later. So, as you make mistake, resolve to do better next time. We all learn more from our mistakes than from our successes—fail forward!
   d. In process two, you identified your primary learning style(s). Now anchor learning by actively utilizing more than your primary learning style (e.g., vicarious learning, visualization, verbalizing instructions or material, etc.).
9. Immersion in the task or activity.
   a. Immerse yourself into the learning context.
   b. The goal is to learn enough to self-correct through observation and practice.
   c. Continue to observe for comprehension and to absorb meaning (vicarious learning/mediation).
   d. Immediately following task engagement reflect on the activity paying attention to key takeaways toward improvements for the next engagement. You will also debrief the activity and receive an objective perspective and feedback to ensure you are on the right track.
10. Seek feedback, establish goals, and persevere.
    a. Feedback should be provided on a continual basis.
    b. Again, feedback provides a mechanism by which corrections can be made (both through external redirection and internal self-direction).
    c. Feedback allows task optimization—an effective balance between challenges and skills. Seeking the optimal balance is a product of persistence and willingness to utilize feedback on an ongoing basis to maximize potential, learn, develop, and set goals.
    d. Debriefing a training session or day of activity facilitates self-efficacy and recall allowing learning to be codified and stored in memory for future retrieval.

    e. Plan for tomorrow's task engagement activity today. Based on lessons learned, feedback and a review of the day's activities prioritize and develop a plan and goals for the next day's work activity. Staying focused and flexible is the key to building task efficacy.

Instruction is most effective when trainers provide structure, consistency, and role modeling. Tenpas offers a task planning model that builds these characteristics into training activities. The model has the trainer enumerating each step (or task element) in the task on a sheet of paper in sequence. Task tools (or assistive technologies) are arranged in front of the consumer, and a process for instruction and learning is determined in a hands-on manner. The trainer then completes an action plan recording in separate columns the tasks, how well the consumer performed, relevant notes, whether the trainer feels the consumer can acquire requisite skills and the training approach utilized (similar to the decision matrix discussed earlier in the text). Tenpas finds that a process like this builds self-efficacy and provides a hard copy of consumer activity (Tenpas, 2003) that may be used to determine consumer informed choice and AT feature-matching throughout the AT service delivery process.

SUMMARY

There are many steps inherent within the assistive technology service delivery process. Implementation may follow the assessment process directly, or it may take some time. Issues such as determining the cost for AT, who should fund needed items, and the elapsed time since assessment can all complicate implementation plans. Implementation is a person-centered process. All aspects of implementation must take into account consumer preferences. Service provision must reflect a well-ordered and planned approach. Immediately following implementation and training, monitoring and feedback with follow-up training as necessary should be provided. Effective training, together with time, effort, and persistence on the part of the consumer should lead to greater self-efficacy with the AT and increased vocational competence overall.

REFERENCES

Bond, G. R., Drake, R. E., Mueser, K. T., & Becker, D. R. (1997). An update on supported employment for people with severe mental illness. *Psychiatric Services, 48*(3), 335–346.

Bruner, J. S. (1966). *Toward a theory of instruction.* Cambridge, MA: Harvard University Press.

Bruner, J. S. (1996). *The culture of education.* Cambridge, MA: Harvard University Press.

Corrigan, P. W., & McCracken, S. G. (2005). Place first, then train: an alternative to the medical model of psychiatric rehabilitation. *Social Work*, *50*(1), 31−39.

Csikszentmihalyi, M. (1990). *Flow: The psychology of optimal experience*. New York: Harper Perennial.

Fadely, D. C. (1987). *Job coaching in supported work programs*. Menomonie, WI: Stout Vocational Rehabilitation Institute, University of Wisconsin-Stout, 54751.

Greeno, J. G., & Collins, A. (2008). Commentary on the final report of the national mathematics advisory panel. educational researcher. *U.S. Department of Education*, *37*(9), 618−623.

Hebb, D. O. (1949). *Organization of behavior: A neuropsychological theory*. New York, NY: John Wiley & Sons.

MacDuff, G. S., Krantz, P. J., & McClannahan, L. E. (2001). Prompts and prompt-fading strategies for people with autism. In C. Maurice, G. Green, & R. M. Foxx (Eds.), *Making a difference: Behavioral intervention for autism* (pp. 37−50). Austin, TX: Pro-Ed.

Rother, M., & Shook, J. (2003). *Learning to see: Value-stream mapping to create value and eliminate muda*. Cambridge, MA: The Lean Institute.

Tenpas, S. (2003). *Job coaching XE "Job coaching" strategies: A handbook for supported employment*. Verona, WI: Attainment Company, Inc.

Wehman, P., Inge, K. J., Revell, W. G., & Brooke, V. A. (2007). *Real work for real pay: Inclusive employment for people with disabilities*. Baltimore, MD: Paul H. Brookes Publishing Co.

Follow-up, follow along, service completion, and outcomes

12

Catherine A. Anderson
Rehabilitation Research and Training Center on Evidence-Based Practice, University of Wisconsin-Madison, Madison, WI United States

FOLLOW-UP, FOLLOW-ALONG, SERVICE COMPLETION, RE-REFERRAL

FOLLOW-UP AND CASE CLOSURE

Once services have been implemented and training completed, the consumer is left to practice and manage assistive technology (AT) either independently or in cooperation with other members of the rehabilitation team or support staff who have been trained alongside them in the use and maintenance of the AT (e.g., personal care provider, job coach, social worker, teacher). There is a window of time the disability-employment professional allows to demonstrate employment stability. AT follow-up services tend to coincide with this window (generally, up to 90 days). Follow-up services provide an opportunity to monitor, measure, and document progress toward consumer goals; both AT and employment goals. The follow-up timeline provides an opportunity for corrective action, remedial training, or additional training. There may be times when assistive technologies require additional reconfiguration and adjustment following brief usage. Likewise, once a consumer has had some time to use the AT, their preferences may change requiring some tweaking to the system or setup.

Unexpected issues may also arise. Typically, when issues occur within the follow-up phase of service delivery, the consumer still has an open case with the disability-employment agency and can continue to obtain necessary AT and other service funding (i.e., additional authorizations for AT services) such as remediation, corrective action, or providing additional training toward job maintenance. Once these issues are resolved, the AT case is then terminated and the consumer and their rehabilitation team are notified, generally with a form letter along including consumer rights information.

Assistive Technology Service Delivery. DOI: https://doi.org/10.1016/B978-0-12-812979-1.00012-6

FOLLOW-ALONG AND RE-REFERRAL

Follow-along services occur over the longer term—generally beginning at 90 days following AT implementation. Disability-employment services may have terminated by this time as well (having attained the employment goal and retained employment for at least 90 days). Follow-along services have the effect of providing data for the AT services program toward evaluating and improving programmatic outcomes. During the follow-along period, many factors may have impacted AT effectiveness. Disabilities can change and become worse (or improve), life changes may impact effectiveness, there may have been a job or job process change, there may be a new boss with a new way of thinking and doing things, there may have been physical changes to the work context (environmental alteridad), etc. A call or visit from the AT professional during follow-along may find a consumer happily working in their chosen vocation or they may find someone who has suffered an exacerbation to a medical condition with the AT purchased for them months ago no longer meeting their current needs.

The follow-along contact by the AT professional or other team member runs through their follow-up/follow-along protocol to determine the need for a new referral for AT services. As a re-referral, however, the referral/funding source would make the determination to send the consumer back through the AT process for a systematic assessment. A referral would be made by the referral/funding source and the client would have their case reassigned, reentering the service delivery process. We will have learned from their first time through, so we can structure things a little differently toward having greater success (being direct about consumer responsibility and clearly discussing next steps). Once services have been provided, problems resolved, the consumer is again stable in their job and satisfied with the services delivered, the case is closed and letters are sent to the consumer and the referral/funding source to this effect. At this point, case files are transferred to team members tasked with ensuring documentation is complete and funding issues have been addressed. Files then go through a quality assurance or outcomes audit and then into storage.

ASSISTIVE TECHNOLOGY AND SERVICE DELIVERY OUTCOMES

An outcome is defined as "something that follows as a result or consequence" (Merriam-Webster, n.d.). Thus, the outcomes of the AT service delivery process must be identified and reported using measures that are meaningful to key stakeholders. The fundamental priorities regarding such outcomes vary among stakeholders. Priorities of outcomes often vary across stakeholders and as an example improved community participation may be the most important outcome for one while enhanced quality of life is the priority for another. Some stakeholders may carry out the roles of more than one stakeholder, for example, providers of AT and rehabilitation services or schools may be one and the same, or users or employers may also be payers.

While all stakeholders emphasize functional capability, most vary in their attention to the person and personal factors, milieu and environment of use, and technology functions and features. Users want products that benefit their performance of activities and participation as well as subjective well-being and quality of life. When an AT product or system meets standards of good design and usability, its use and the outcomes of its use depend heavily on initial expectations of benefit, involvement in product and feature selection, and adequate training for use (Scherer & Federici, 2017).

Apart from family members having an interest in optimum user outcomes, AT products can also bring positive outcomes in their lives. For example, care burden may be reduced when a user can control his or her environment through a remote control, or stress and conflicts may be reduced when a child with attention deficit hyperactivity disorder (ADHD) uses a time management device.

Providers of AT need to emphasize consumer satisfaction with the devices and services received while containing costs. Prescribers and counselors are licensed or certified to improve the functional situation of the people they serve. They enhance quality of life and subjective well-being as well as social participation (Scherer & Federici, 2017). Their assessment is key to identify individually appropriate AT solutions. Users may be satisfied with the services, have the necessary funding, have received a product that is usable, looks good, functions well, and meets all safety standards, and have helped them achieve functional gain—but if it is a hassle to use, set up, and maintain the solution, if it doesn't fit with their needs, preferences, or lifestyle, if they feel self-conscious using it, insecure with use even though it is safe, if they are socially and physically and emotionally uncomfortable with use, then they are not realizing benefit from use and will not use it. Ultimately, it is the user experience and realization of benefits that determine whether AT products are used (Scherer & Federici, 2017).

WHY DO WE NEED OUTCOMES MEASUREMENTS IN ASSISTIVE TECHNOLOGY SERVICE DELIVERY?

Outcomes are typically considered to be the goals we want to achieve in providing AT services involving assessment, training, or implementation. Funders, individuals with disabilities, and family members have every right to ask that practitioners show measureable results that demonstrate progress toward the stated goals and address the question of which AT solution best meets a client's needs, or in other words, did the service achieve its anticipated objective (Satterfield, 2016)? It helps us not only identify the intended goals but also produce meaningful results in terms of efficacy of the service provided. As the evidence-based practice movement gains momentum across rehabilitation and healthcare professions globally, the ability for AT practitioners to clearly define and report outcome measures connected to service delivery will be of great benefit to consumers, funders, and the field in general.

While AT evaluation and training are clearly important elements, AT researchers strongly recommend adding outcomes measurement as a critical third component to service delivery measuring the impact of an AT intervention (Edyburn, 2011; Scherer, 1996). In the educational system, there is broad agreement that use of AT positively impacts students with disabilities; however, research-based outcome measurement tools needed to document efficacy are not implemented consistently (Edyburn, 2015). This will likely be changing in the coming years as more tools become available and funding sources require that the data be collected and reported. It is logical to expect this as many systems including education, vocational rehabilitation, and health care place increasing emphasis on directing funding toward services that are both efficacious (proven to work) and cost-effective (reasonably priced).

When identifying the outcomes to be measured, it is important to include multiple stakeholders in the discussion as differing viewpoints and goals may be present. As an example, funders may be focused on cost and efficacy while the consumer may have other considerations such as reduced stigma or enhanced community involvement, including employment. It is important for the AT practitioner to gather this input from multiple stakeholders and integrate it into the service plan to ensure proper measurement across multiple outcome domains.

A global framework. The International Classification of Functioning, Disability and Health (ICF), developed by the World Health Organization, provides a meaningful framework in considering how best to identify and design outcome measures related to AT. The overarching goals of the ICF are to provide a standard language and structure for describing and measuring health measures as well as provide a "meaningful and practical system that can be used by various consumers for healthy policy, quality assurance, and outcome evaluation in different cultures" (World Health Organization, 2001). The ICF framework acknowledges functioning and disability within the context of contextual factors, and integrates specific components, domains, and constructs. While some have argued that the ICF framework may not be the best fit for AT, others strongly advocate for its potential as a comprehensive and culturally responsive option and recommend further research to develop indicators and practical instruments specific to measuring AT outcomes.

IDENTIFYING AND MEASURING ASSISTIVE TECHNOLOGY ASSESSMENT, IMPLEMENTATION, AND OUTCOMES

Over the last 20 years, a number of projects have specifically addressed issues involved with identifying and mapping AT outcomes. It is not accidental that the advent of these efforts coincides with the emergence of the evidence-based practice movement in health care, which has consistently expanded across public health, social work, and rehabilitation professions. Before outcomes can be measured and standardized, they must be identified. As noted earlier, inclusion of a number of key stakeholders including AT users, funders, family members, educators, policy makers, and community members, using participatory methods, can strengthen the process and ensure that varying perspectives and needs are addressed.

An example of a collaboratively developed model used to identify outcomes and "best practices" in educational settings is the Quality Indicators for AT (QIAT) (QIAT Consortium Leadership Team, 2000). While originally formed nearly 20 years ago, the QIAT leadership continues to provide guidance on the provision of AT services in educational settings, and many of their resources and tools, including the quality measurement matrices noted below, are available free of charge at http://qiat.org/indicators.html. This entity has mapped out quality indicators across the key phases of (1) AT assessment or evaluation, (2) AT implementation, and (3) AT outcomes (referred to as effectiveness by QIAT), and recently updated these in 2015 to reflect contemporary practice. The quality indicators are measured along five-point Likert scales with 1 = unacceptable practice and 5 = promising practice.

Quality indicators for assessment of AT needs. The following quality indicators have been defined by QIAT and can be measured during the *assessment or evaluation* phase of AT service provision.

1. Procedures for all aspects of AT assessment are clearly defined and consistently applied.
2. AT assessments are conducted by a team with the collective knowledge and skills needed to determine possible AT solutions that address the needs and abilities of the student, demands of the customary environments, educational goals, and related activities.
3. All AT assessments include a functional assessment in the student's customary environments, such as the classroom, lunchroom, playground, home, community setting, or work place.
4. AT assessments, including needed trials, are completed within reasonable timelines.
5. Recommendations from AT assessments are based on data about the student, environment, and tasks.
6. The assessment provides the educational team with clearly documented recommendations that guide decisions about the selection, acquisition, and use of AT devices and services.
7. AT needs are reassessed any time changes in the student, the environments, and/or tasks result in the student's needs not being met with current devices and/or services.

Quality indicators for AT implementation. The following quality indicators have been defined by QIAT and can be measured during the *implementation* phase of AT service provision.

1. AT implementation proceeds according to a collaboratively developed plan.
2. AT is integrated into the curriculum and daily activities of the student across environments.

3. Persons supporting the student across all environments in which the AT is expected to be used share responsibility for implementation of the plan.
4. Persons supporting the student provide opportunities for the student to use a variety of strategies, including AT, and to learn which strategies are most effective for particular circumstances and tasks.
5. Training for the student, family, and staff is an integral part of implementation.
6. AT implementation is initially based on assessment data and is adjusted based on performance data.
7. AT implementation includes management and maintenance of equipment and materials.

Quality indicators for evaluation of the effectiveness of AT. The following quality indicators have been defined by QIAT and can be used to measure the *effectiveness* of AT services provision.

1. Team members share clearly defined responsibilities to ensure that data are collected, evaluated, and interpreted by capable and credible team members.
2. Data are collected on specific student achievement that has been identified by the team and is related to one or more goals.
3. Evaluation of effectiveness includes the quantitative and qualitative measurement of changes in the student's performance, achievement, or function.
4. Effectiveness is evaluated across environments including naturally occurring opportunities as well as structured activities.
5. Data are collected to provide teams with a means for analyzing student achievement and identifying supports and barriers that influence AT use to determine what changes, if any, are needed.
6. Changes are made in the student's AT services and educational program when evaluation data indicate that such changes are needed to improve student achievement.
7. Evaluation of effectiveness is a dynamic, responsive, ongoing process that is reviewed periodically.

While the QIAT matrices present one resource for identifying and measuring outcomes across the phases of AT delivery, additional options are presented for consideration. As an example, Lenker and Paquet (2004) recommend use of a model including (1) AT usability as defined by duration, frequency, environments, context, and tasks and (2) quality of life involving health and well-being, quality of social relationships, and ability to perform in social roles as key outcome indicators. Additionally, Fuhrer, Jutai, Scherer, and DeRuyter (2003) suggest use of a model for measuring AT outcomes inclusive of both objective and subjective measures involving (1) data that address stakeholder concerns, (2) give primary consideration to the goals and needs of the individual AT user, (3) address the need for common definitions such as the ICF, and (4) provide reflectiveness regarding mediating and moderating factors.

THE ROLE OF ACCREDITATION IN QUALITY ASSURANCE

Accreditation bodies also serve as a resource in providing guidance regarding performance management and quality standards that organizations employing rehabilitation professionals should strive for. Accreditation organizations typically embrace a collaborative approach to the identification, mapping, and updating of key standards used to guide and measure quality service delivery processes and results. CARF International is an established and globally recognized accreditation body with offices in Europe, Canada, and the United States, whose mission is to "promote the quality, value, and optimal outcomes of services through a consultative accreditation process and continuous improvement of services that center on enhancing the lives of persons served" (see CARF International http://www. carf.org/home). The CARF Advisory Council involves an interdisciplinary array of professionals, with specific subject matter expertise including considerable input from the field, used to develop and update the recommended standards. They provide comprehensive standards regarding AT supports and services within their Employment and Community Services domain. CARF International identifies the following examples of quality results (outcomes) of AT assessment and implementation services:

1. Increased independence.
2. Increased community access.
3. Increased participation of the person in the community.
4. Increased employment options.
5. Increased wages.
6. A flexible, interactive process that involves the person served.
7. Individualized, appropriate accommodations.
8. Decreased family or caregiver support.
9. Timely services and reports.

CARF delineates quality indicators throughout their standards but offers a binary approach (yes or no) rather than a scaled approach such as Likert to their measurement. CARF notes that AT can be used to enhance lives and independence across a number of areas including education, employment, communication, community involvement, housing, recreation, public facilities, transportation, telecommunications, activities of daily living/independent living, and others. A number of "intent statements" are highlighted across the standards placing specific focus on the values, principles, and quality indicators involved. As an example, within the Assistive Technology Standards section, specific attention is given to persons and/or families served participated in making informed decisions about their AT services including

1. Expected results (outcomes) of services
2. How the design of services meets identified needs
3. Off-the-shelf technology resources, as appropriate
4. How services will be delivered

5. Expected timeline of services
6. Possible alternatives for services
7. How results will be evaluated
8. Full disclosure to persons served and funders about technology benefits, maintenance, expected costs, expected responsibilities, and technology changes
9. Other aspects about the service design, as requested (CARF, 2017).

The required individual AT service plan components identified by CARF includes (1) identification of functional limitations to opportunities of the person served, (2) addresses potential for accommodations, (3) integrated accommodations into the current employment situation, if applicable, (4) addresses previous AT services, if applicable, (5) addresses the dynamic nature of the disability, as appropriate, (6) addresses anticipated changes in employment, environment, or living situation, as appropriate, and (7) addresses healthy and safety risks, as appropriate.

While CARF does not specifically use the term "outcomes," they frequently refer to results throughout their standards. Additionally, they specify that when a person exits AT services, the summary report must contain

1. Description of the AT services provided and approximate training time,
2. Identification of potential future AT needs and recommended implementation plan, and
3. Maintenance, troubleshooting, and repair sources.

CONCLUSION

In conclusion, the development of consistent, systematic outcomes measures for use across AT services remains in its infancy. However, practitioners are encouraged to research existing options for use in individual practice through sources such as QIAT as well as the Rehabilitation Engineering and Assistive Technology Society of North America as they become available. It is anticipated that an increasing number of high-quality outcomes measures will be identified and developed in the coming years, which presents an exceptional opportunity for AT practitioners to engage in implementing these in practice. In all likelihood, consistent outcome measures will involve quality indicators including full inclusion of service consumers, timeliness of service delivery, transparency throughout service delivery, improvements in function, health, and performance, and cost-effectiveness. Ultimately, use of robust outcomes measures will drive excellence in practice and provide an opportunity for professionals to demonstrate and document that AT service delivery results in meaningful and measureable outcomes for individuals with disabilities.

REFERENCES

CARF. (2017). Employment and community services standards manual: Assistive technology supports and services.

Edyburn, D. L. (2011). Some of the best: Advances in special education technology research. *Closing the Gap, 29*(6), 30–33.

Edyburn, D. L. (2015). Expanding the use of assistive technology while mindful of the need to understand efficacy. In D. L. Edyburn (Ed.), *Advances in special education technology, Volume 1: Efficacy of assistive technology interventions* (pp. 1–12). Bingler, UK: Emerald Group, Publishing Limited.

Fuhrer, M. J., Jutai, J. W., Scherer, M. J., & DeRuyter, F. (2003). A framework for the conceptual modeling of assistive technology outcomes. *Disability and Rehabilitation, 25*, 1243–1251.

Lenker, J. A., & Paquet, V. L. (2004). A new conceptual model for assistive technology outcomes research and practice. *Assistive Technology, 16*, 1–10.

Merriam-Webster. n.d. "Outcome." Available at: https://www.merriam-webster.com/dictionary/outcome (Accessed 7 July 2018).

QIAT Consortium Leadership Team. (2000). Quality indicators for assistive technology services in school settings. *Journal of Special Education Technology, 15*(4), 25–36.

Satterfield, B. (2016). History of Assistive Technology outcomes in education. *Assistive Technology Outcomes and Benefits, 10*(1), 1–18.

Scherer, M. & Federici, S. (2017). *Assistive technology assessment handbook* (2nd ed.). *Preface to second edition.* Boca Raton, FL: CRC Press.

Scherer, M. J. (1996). Outcomes of assistive technology use on quality of life. *Disability and Rehabilitation, 18*(9), 439–448.

World Health Organization. (2001). *International classification of functioning, disability and health: ICF.* Geneva, Switzerland: World Health Organization.

Assistive Technology Toolkit

3

In Part 3 of this text, we offer an assistive technology "toolkit." The four chapters which comprise the toolkit offer perspectives on service delivery which are of growing interest and importance within disability-employment service delivery (Chan, Tarvydas, Blalock, Strauser, & Atkins, 2009) generally and assistive technology service delivery specifically. In Chapter 13, *Evidence-Based Practice*, we provide an overview of evidence-based practice (EBP)—what EBP is, how knowledge translation informs EBP, and why EBP is important to service delivery. Chapter 14, *Consideration of Underserved Groups and Lower Resourced Environments* provides perspective on the intersection of poverty, health, disability, and multicultural competence. In Chapter 15, *On Technology Abandonment or Discontinuance*, we offer our thoughts regarding the factors which may impact a consumer's decision to stop utilizing the assistive technology that has been provided to them. Finally, Chapter 16, *Assistive Technology Techniques, Tools, and Tips* offers several fundamental tools which may be used to great effect in the assistive technology service delivery process. These include the Matching Person and Technology Model, case management types, and case noting in disability-employment service delivery.

The *Assistive Technology Toolkit* brings together resources which inform the assistive technology service delivery process. The chapters in the toolkit offer perspectives gaining attention and valued by disability-employment professionals who seek to educate, empower, and employ people with disabilities and improve outcomes (Project E3, 2018).

REFERENCES

Chan, F., Tarvydas, V., Blalock, K., Strauser, D., & Atkins, B. J. (2009). Unifying and elevating rehabilitation counseling through model-driven, diversity-sensitive evidence-based practice. *Rehabilitation Counseling Bulletin, 52*, 114—119.

Project E3. (2018). *About project E3*. Vocational Rehabilitation Technical Assistance Center for Targeted Communities. Retrieved from ⟨https://projecte3.com/about/⟩.

Evidence-based practice

13

Catherine A. Anderson

Rehabilitation Research and Training Center on Evidence-Based Practice, University of Wisconsin-Madison, Madison, WI, United States

OVERVIEW

What is evidence-based practice (EBP) and why should service providers care? For some, terms such as "evidence" and "research" may sound unappealing given that their focus as practitioners is on helping people, not necessarily concern about data and "scientific" approaches. However, interest in providing high-quality services that are known to work, at a reasonable cost, using practical approaches is exactly *why* developing a better understanding of what EBP *is* and *is not* is becoming increasingly important.

Advocates for EBP support that all rehabilitation and allied health professionals should have an interest in delivering the best possible services to their consumers, based on the best clinical practices available from the research evidence, to the greatest extent possible (Chan, Tarvydas, Blalock, Strauser, & Atkins, 2009). Funders of services are also increasingly attentive to the application of EBP and it is likely that rehabilitation professionals and others will be asked to better integrate research evidence in their professional decision-making processes in the near future (Chan, Miller, Pruett, Lee, & Chou, 2003; Schlosser, 2006). In this chapter, we will cover the following information to better understand concepts related to EBP and discuss how to apply them in practice:

- Defining EBP
- The evidence-based continuum
- Assistive technology (AT) service and practice on the evidence-based continuum
- Implementation of EBP: benefits and challenges
- Effective integration of EBP into service delivery

DEFINING EVIDENCE-BASED PRACTICE

The EBP movement emerged first in medicine in the late 20th century, followed closely by the public health and mental health professions, and is now an international effort in which evidence serves as a guiding force in defining practice and

Assistive Technology Service Delivery. DOI: https://doi.org/10.1016/B978-0-12-812979-1.00013-8

policy (Evidence-Based Medicine Working Group, 1992; Tucker & Reed, 2008). Since its inception, professionals including physicians, nurses, physical therapists, occupational therapists, speech pathologists, rehabilitation counselors, and others have become increasingly interested in the available evidence to support decision-making in practice (Bezyak, Kubota, & Rosenthal, 2010; Sackett, Rosenberg, Gray, Haynes, & Richardson, 1996). While the evidence base related to AT interventions is relatively small at this point, it is anticipated that the EBP movement will become increasingly central in this field as well. Importantly, EBP is not about using research to replace clinical or professional expertise, but rather integrating research-based knowledge and practices into professional decision-making, care, and service delivery.

The hallmark features of EBP are generally referred to as the "Three Es" meaning that the practice has been determined to be effective, efficient, and efficacious (Lui & Anderson, 2013). *Effectiveness* refers to the practices' usefulness or functional application while *efficiency* refers to cost-effectiveness, or the ability to provide the service at a reasonable expense. And to be identified as *efficacious*, the practice must have rigorous research demonstrating statistical and clinical significance to support that it works. Efficiency in particular is an important consideration as this can be a driving condition for policy makers, administrators, and funding sources such as insurance companies, public programs, and others. A service can be identified as effective or easy to provide and have demonstrated efficacy through the research, but if it is expensive to administer, it is unlikely to be adopted into practice and available to the broader population. That said, it is also important for funding sources including private insurance and public programs not let the need for efficiency outweigh the other factors.

THE EVIDENCE-BASED CONTINUUM

In defining EBP, it is helpful to consider it in the context of a continuum. **Emerging** practices offer informed suggestions about what works and does not work but doesn't yet have evaluation data to demonstrate its effectiveness (Puddy & Wilkins, 2011; Twyman & Sota, 2008). **Promising** practices may include elements of emerging practice elements and add programmatic quantitative data that demonstrates positive outcomes but may not yet contain research data that supports replication of the service or intervention (DePalma, 2002). And finally, **EBPs** are those with demonstrated statistical and clinical significance (Driever, 2002). While use of EBP is preferable, it may not yet be available in which case it is advisable to utilize promising and emerging practices as needed (Fig. 13.1).

HIERARCHICAL LEVELS OF EVIDENCE

Complexities occur when trying to define "best evidence" as indicated by an ongoing debate in the literature regarding quantity and quality of research needed and lack of consensus on how to best apply research evidence and define the effectiveness of approaches (Detrich, Keyworth, & States, 2007; Odom et al., 2005; Tannenbaum, 2005). However, the hierarchy delineating the levels of evidence, as consistently defined by Holm (2000), Gray (1997), and Nathan and Gorman (1998), is commonly recognized and includes five levels. Level one is considered the strongest evidence and level five is considered the least robust, but still relevant for practice (Fig. 13.2).

Level 1: Strong evidence from at least one systematic review of multiple well-designed randomized controlled trials.

Level 2: Strong evidence from at least one properly designed randomized controlled trials of appropriate size.

Level 3: Evidence from well-designed trials without randomization, single group pre—post, cohort, time series, or matched case-controlled studies.

Level 4: Evidence from well-designed nonexperimental studies from more than one center or research group.

Level 5: Evidence from opinions of respected authorities, based on clinical evidence, descriptive studies, or reports of expert committees.

While the emphasis remains on using the most rigorous evidence possible, there is flexibility based on an understanding that variability exists in the amount and type of evidence present and subsequently available for application in practice.

FIGURE 13.1

The emerging, promising, and evidence-based continuum.

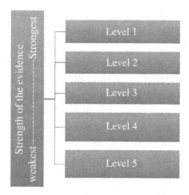

FIGURE 13.2

Hierarchy depicting levels of evidence-based practice based on strength of the evidence supporting the practice.

KNOWLEDGE TRANSLATION: DEMYSTIFYING RESEARCH FOR PRACTICE

Knowledge translation (KT) is generally defined as "the multidimensional, active process of ensuring that new knowledge gained through the course of research ultimately improves the lives of people with disabilities and furthers their participation in society" (Sudsawad, 2007). KT strategies are used to enhance the adoption of EBPs in the field by providing practical interpretation of research-findings and making them available in user-friendly formats to help bridge the gap between research and practice. The gap between research and practice exists and scholars note that (1) it hinders the adoption of EBP in practice, (2) must be addressed in order for practitioners to understand and use EBP in practice, (3) resulting in improved outcomes with consumers (Anderson, Matthews, Lui, & Nierenhausen, 2014; Fleming, Del Valle, Kim, & Leahy, 2013; Murray, 2009).

How can we effectively bridge the gap between research and practice? Developing materials and products is important, but as Jacobson, Butterill, and Goering (2003) note, KT activities must focus as much on **process** as on **product** and "emphasize crucial elements of reciprocity and exchange between the producers and users of knowledge." Another way to think about this is to consider KT as an interactive, bidirectional approach involving functional information sharing between researchers and practitioners. This contrasts with the more traditional unidirectional approach of researchers putting forth findings and publications in a passive transfer of information without seeking input, feedback, and ongoing dialogue from and with the field. Traditional methods make it challenging for those not working in university settings to access the information. Additionally, while the information is generally presented in formats that adhere to established standards in the scholarly community, this may have little relevance for application in the field.

KT framework. Within KT frameworks, researchers and practitioners work collaboratively to ensure that new EBPs are informed with contextual knowledge and practical experience regarding how to share the information and apply it in practice (Kerner, 2006; Lui, Anderson, Matthews, Nierenhausen, & Schlegelmilch, 2014). Of the various KT frameworks available, the knowledge-to-action (KTA) framework proposed by Graham et al. (2006) is gaining popularity as a good fit within the rehabilitation context. The KTA framework involves a knowledge creation phase involving inquiry, synthesis, and tool or product development, followed by an action phase involving knowledge application with clear feedback loops to assess and address contextual issues, evaluate use and outcomes, and determine sustainability of a practice (Graham et al., 2006). Information gathered through implementation is used to continually inform improvements and/or modifications to practice and process (Fig. 13.3).

KT strategies. When asked about preferred formats for acquiring and sharing information about EBPs, rehabilitation professionals clearly noted a preference for informal, nonacademic formats that can be read and processed in a relatively

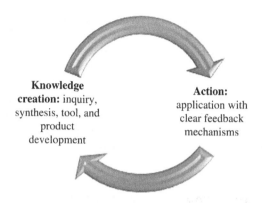

Knowledge creation: inquiry, synthesis, tool, and product development

Action: application with clear feedback mechanisms

FIGURE 13.3

Modified depiction of the knowledge-to-action knowledge translation framework.

short amount of time (Anderson et al., 2014). Specific examples include brief summaries of findings in user-friendly formats such as plain language summaries, face-to-face training, online asynchronous (noninteractive) training, and training conferences.

Reputable professional websites, facilitated online forums, and communities of practice (CoP) provide opportunities to openly discuss and exchange knowledge and resources can be inexpensive sources of helpful information. CoP in particular can serve as excellent resources for professionals to connect, converse, and support each other in identifying resources as well as problem solving specific situations (Wenger, 1998; Wenger, McDermott, & Snyder, 2002). As online and technological options continue to expand, expect to see EBP information become more readily available through easily accessible channels.

ASSISTIVE TECHNOLOGY PRACTICE AND THE EBP CONTINUUM

While the integration of EBP is relatively new to the fields of AT and rehabilitation counseling, the consideration of AT has been a mandate in the development of every student with a disability in the United States for nearly 20 years [20 U.S. C. 1401 § 614(B)(v)] and the use of AT is believed to enhance educational outcomes for students [§ 616(a)(2)(A)]. Furthermore, the use of AT in school settings presents the opportunity to not only enhance or improve student access to education and learning but also improve performance in full community inclusion including competitive integrated employment, recreation, independent living, and social opportunities (Peterson-Karlan & Parette, 2007).

In order to initiate the implementation of new research evidence into practice, it is helpful to first assess the level of understanding current rehabilitation professionals have regarding EBP. A recent study measured certified rehabilitation counselors' self-reported (1) attitudes and beliefs about EBP; (2) education, knowledge, and skills related to obtaining and evaluating evidence; (3) attention to the literature relevant to practice; (4) access to and availability of information; and (5) perceptions of the barriers to EBP (Bezyak et al., 2010). Overall, attitudes were positive regarding interest in learning skills needed to effectively incorporate evidence into rehabilitation counseling practice. These findings are consistent with those shared by Anderson et al. (2014) with professionals reporting interest in learning about and applying EBP in the field.

FOUR STEP PROCESS FOR IMPLEMENTING EBP

EBP is typically described as a process involving four specific steps (Chronister, Chan, Cardoso, Lynch, & Rosenthal, 2008; DePalma, 2002; Parkes, Hyde, Deeks, & Milne, 2001). The steps are designed to be implemented sequentially as they provide a logical flow to identifying whether there is an established practice for use with a particular consumer. Chan et al. (2010) provide a clear overview of how to conceptualize and apply the four-step process in practice.

Step 1. Formulate well-defined, answerable questions.
Step 2. Seek the best evidence available to answer the questions.
Step 3. Critically appraise the evidence.
Step 4. Apply the evidence to the individual client.

Step 1. When formulating questions in Step 1, it is important to ask both **general questions** related to background setting and context as well as **"foreground" questions**, or those that focus on details about a specific case within that context (Walker, Seay, Solomon, & Spring, 2006). Walker et al. (2006) provide examples of general questions the rehabilitation professional may ask including

- What processes and/or techniques make a specific intervention work?
- For whom is the intervention most effective?
- Are certain interventions and/or programs better for certain persons?
- Who should receive a specific intervention or program? When? And for how long?

The PICO format involves answering questions related to (P): patient group, (I): intervention, (C): comparison group, and (O): outcome measures is recommended for use in forming foreground questions to focus on issues specific to the case (Walker et al., 2006). The tool provides professionals with a framework that is easily applied in practice and provides a consistent and objective approach to formulating questions across populations and consumer groups. By asking both background (general) and specific (foreground) questions, it provides a basis for identifying research-based practice that fits with the context of the individual case.

Scenario for applying PICO. Let's walk through an example of how to apply Step 1. Your client is a 20-year-old Caucasian college student with low vision who is frustrated with his current hand-held magnifier and would like to explore online/electronic alternatives. Background questions may include

- What are the most effective AT interventions for low vision?
- Are online tools such as apps effective interventions with low vision?
- Are there any significant risks associated with using online tools?

PICO, or foreground information:

(P): a young adult male with low vision ("patient," although we typically refer to individual clients, consumers, or students in rehabilitation),
(I): is there any evidence that online/electronic tools (intervention),
(C): are superior to hand-held magnifiers (comparison),
(O): in improving access to information and reducing client frustration (outcomes)?

Answering these questions provides a good foundation for case development and offers direction for determining what evidence to look for and where to search for the best evidence (Chan et al., 2010).

Step 2. Following formulation of background and foreground questions, the next step involves searching for EBP s or interventions that may apply to the case. While searching scholarly and academic databases is recommended, many practitioners may not have access to these resources unless employed through a university or teaching hospital. Alternatively, there are four highly regarded websites that professionals can use to search for evidence-based medical, rehabilitation, and behavioral science interventions and practices. The Cochrane Collaboration and the Campbell Collaboration (2018) provide an international perspective and access to resources that provide a global perspective. The Agency for Healthcare Research and Quality (2018) and the American Congress of Rehabilitation Medicine (2018) are also reputable resources to use (Fig. 13.4).

When searching these websites, Chan and colleagues (2010) recommend entering key words related to the issue at hand along with the terms systematic review or metaanalysis. If there is relevant evidence available, searching in this manner will help prioritize the strongest evidence as it denotes multiple research studies connected to the issue at hand.

Step 3. As noted in Step 2, using systematic reviews when available can be helpful in determining strength of the evidence and can also save time rather than the professional reviewing multiple studies individually. A systematic review, as described by both Chan and colleagues (2010) and Schlosser (2006), is "a thorough, comprehensive, and explicit process of collecting, reviewing and presenting all available evidence ... abstracting data in a standardized format and providing a summary of the evidence from the primary sources." Metaanalyses, also referred to as "a study of the studies," and randomized controlled trials using experimental design serve well in informing the efficacy of a particular service or intervention.

The Cochrane Collaboration
www.cochrane.org

The Campbell Collaboration
www.campbellcollaboration.org

Agency for Healthcare Research and Quality
www.ahrq.gov

American Congress of Rehabilitation Medicine
http://www.acrm.org/consumer_professional/Evidence_Based_Practice.cfm

FIGURE 13.4

Examples of reputable sources of evidence-based practices.

Step 4. Once the research evidence has been gathered and reviewed, the practitioner then decides whether and how best to integrate it into the client's plan by weighing the significance of the evidence along with the contextual information using his or her professional expertise and judgment.

While it may be challenging to find many Level 1 or 2 AT practices available for implementation at this time, there may be options that meet the criteria at other levels and given the context of the situation, may be very appropriate to apply in practice.

IMPLEMENTATION OF EVIDENCE-BASED PRACTICES: BENEFITS AND CHALLENGES

While the EBP movement is not without its controversy and critics, it presents a window of opportunity for the rehabilitation professions to promote and support a systematic agenda for theory-driven research. The use of scientific evidence derived from theory-driven research to information rehabilitation counseling practices has the potential to contribute to higher employment rates and better quality of employment outcomes for people with disabilities (Chan et al., 2010). Benefits and challenges exist in implementing EBP and it is important to have a general understanding of the concepts.

BENEFITS

The primary benefit and intended outcome of using EBP is to improve consumer outcomes and support individuals in achieving their desired level of comfort and autonomy in the workplace, community, home, school, and leisure. The ability to achieve better outcomes consistently, in a practical manner, at a lower cost to individuals and society is desirable. There is tremendous potential to develop, implement, and study the efficiency, effectiveness, and efficacy of EBP across a wide variety of AT domains.

BARRIERS

One of the most evident barriers to implementing EBP appears to be a lack of academic preparation to use evidence in practice. Additional barriers identified by rehabilitation counselors included insufficient time and difficulty in applying research results to counseling practice and/or individual clients (Chan et al., 2010). The latter may indicate a lack of effective KT strategies in use. Barriers can also be symptomatic of organizational environments that have not yet identified the need to integrate evidence into practice (Winch, Henderson, & Creedy, 2005; Yousefi-Nooraie, Dobbins, & Marin, 2014). And a weak theoretical base and uncertain research quality are two major barriers to the successful implementation of rehabilitation research into practice (Bellini & Rumrill, 2002; Chwalisz & Chan, 2008).

Limited EBP in AT. As is the case with many professions, the EBP developed has not kept pace with the increasing interest and need in the field. As of 2012, researchers found 44 research-based articles related to AT (Brandt & Alwin, 2012). No articles about the effectiveness of AT were found, and a specific lack of evidence on AT for children and younger individuals as well as a lack of evidence on the outcomes of AT on daily activities and participation in the workforce, home, and recreation or leisure was mentioned. A similar dearth of EBP specific to AT was found by conducting a search through the Cochrane Collaboration (2018). Van der Roest, Wenborn, Pastink, Dröes, and Orrell (2017) sought evidence on the use of AT for memory support in dementia but found that while some small studies had tested the effectiveness of selected AT devices, the methods used were not deemed to be high enough quality to meet the review criteria. Therefore, the authors could not determine whether AT is effective in supporting people with dementia to manage their memory problems. Additionally, Thomas, Barker, Rubin, and Dahlmann-Noor (2015) noted the lack of research conducted on the use of AT to assist children and young people with low vision, and subsequently the impact on outcomes in independent learning and quality of life. They emphasize the need for study in this area and encourage the dissemination of outcomes in a fashion relevant not only to researchers but also more importantly to families and teachers.

While there may not be a considerable foundation of strong EBP specific to AT yet, it is highly likely that the research base in this area will grow in the coming years. Given that those practicing as assistive technology professionals (ATP) represent diversity in educational and experiential backgrounds, expect to see the evidence emerge through rehabilitation engineering, occupational therapy, rehabilitation counseling, and other similar professions.

EFFECTIVE INTEGRATION OF EVIDENCE-BASED PRACTICE INTO SERVICE DELIVERY

Prior to fully initiating the use of EBP in a professional field, it can be valuable to assess the readiness of practitioners to learn and adopt the new mindset and practices. Anderson et al. (2014) found that vocational rehabilitation (VR) professionals are interested in (1) acquiring new knowledge that promotes EBP in VR, (2) applying new knowledge that may affect employment outcomes for people with disabilities, and (3) sharing this new knowledge with colleagues. Another group of researchers studied vocational rehabilitation professionals' level of understanding of EBP related to (1) perceived self-efficacy, or confidence in identifying and using the practices; (2) outcome expectancy, or what benefit it might contribute to client outcomes; and (3) readiness to use evidence in current practice (Tansey, Bezyak, Chan, Leahy, & Lui, 2014). The findings support outcome expectancy in significantly predicting whether professionals are ready to adopt the use of EBP, while agency and personal barriers may negatively impact readiness to use EBP.

CONCLUSION

Overall, the development of EBP for use by ATP holds tremendous potential moving forward. The current lack of strong evidence to support the work should not dissuade practitioners from seeking research-based approaches. The likelihood of a variety of meaningful EBP s emerging in the coming years is high and becoming familiar with the hierarchy and relevant resources to access the material will be beneficial to professionals and the individuals they serve.

REFERENCES

Agency for Healthcare Research and Quality, 2018. ⟨www.ahrq.gov⟩.

American Congress of Rehabilitation Medicine, 2018. ⟨http://www.acrm.org/consumer_professional/Evidence_Based_Practice.cfm⟩.

Anderson, C. A., Matthews, P., Lui, J., & Nierenhausen, E. (2014). Engaging vocational rehabilitation counselors in knowledge translation processes: A participatory action approach. *Rehabilitation Counselors and Educators Journal, 7*(2), 6–16.

Bellini, J., & Rumrill, P. (2002). Contemporary insights in the philosophy of science: Implications for rehabilitation counseling research. *Rehabilitation Education, 16*, 115–134.

Bezyak, J. L., Kubota, C., & Rosenthal, D. (2010). Evidence-based practice in rehabilitation counseling: Perceptions and practices. *Rehabilitation Education, 24*(1–2), 85–96.

Brandt, A., & Alwin, J. (2012). Assistive technology outcomes research: Contributions to evidence-based assistive technology practice. *Technology and Disability, 24*, 5–7.

Chan, F., Bezyak, J., Ramirez, M. R., Chiu, C. Y., Sung, C., & Fujikawa, M. (2010). Concepts, challenges, barriers, and opportunities related to evidence-based practice in rehabilitation counseling. *Rehabilitation Education, 24*(3−4), 179−190.

Chan, F., Miller, S., Pruett, S., Lee, G., & Chou, C. (2003). Rehabilitation research. In D. Maki, & T. Riggar (Eds.), *Handbook of rehabilitation counseling* (pp. 159−170). New York, NY: Springer.

Chan, F., Tarvydas, V., Blalock, K., Strauser, D., & Atkins, B. J. (2009). Unifying and elevating rehabilitation counseling through model-driven, diversity-sensitive evidence-based practice. *Rehabilitation Counseling Bulletin, 52,* 114−119.

Chronister, J. A., Chan, F., Cardoso, E., Lynch, R. T., & Rosenthal, D. (2008). The evidence-based practice movement in healthcare: Implication for rehabilitation. *Journal of Rehabilitation, 74*(2), 6−15.

Chwalisz, K., & Chan, F. (2008). Methodological advances and issues in rehabilitation psychology: Moving forward on the cutting edge. *Rehabilitation Psychology, 53,* 251−253.

DePalma, J. A. (2002). Proposing an evidence-based policy process. *Nursing Administration Quarterly, 26*(4), 55−61.

Detrich, R., Keyworth, R., & States, J. (2007, February). *Evidence-based education: It isn't as simple as you might think.* In: *Paper presented at the 25th Annual Western Regional Association of Behavior Analysis.* Burlingame, CA.

Driever, M. J. (2002). Are evidence-based practice and best practice the same?. *Western Journal of Nursing Research, 24*(5), 591−597.

Evidence-Based Medicine Working Group. (1992). Evidence-based medicine. A new approach to teaching the practice of medicine. *Journal of American Medical Association, 268,* 2420−2425. Available from https://doi.org/10.1001/jama.268.17.2420, PMID 1404801.

Fleming, A. R., Del Valle, R., Kim, M., & Leahy, M. (2013). Best practice models of effective vocational rehabilitation service delivery in the public rehabilitation program: A review and synthesis of the empirical literature. *Rehabilitation Counseling Bulletin, 56*(3), 146−159.

Graham, I. D., Logan, J., Harrison, M. B., Straus, S. E., Tetroe, J., Caswell, W., & Robinson, N. (2006). Lost in knowledge translation: Time for a map? *The Journal of Continuing Education in the Health Professions, 26,* 13−24.

Gray, J. A. M. (1997). *Evidence-based healthcare: How to make health policy and management decisions.* New York, NY: Churchill Livingston.

Holm, M. B. (2000). Our mandate for the new millennium: Evidence-based practice. *American Journal of Occupational Therapy, 54,* 575−585.

Jacobson, N., Butterill, D., & Goering, P. (2003). Development of a framework for knowledge translation: Understanding user context. *Journals of Health Sciences & Research Policy, 8,* 94−99.

Kerner, J. F. (2006). Knowledge translation version knowledge integration: A funder's perspective. *The Journal of Continuing Education in the Health Professions, 26*(1), 72−80.

Lui, J. & Anderson, C. (2013). *Promoting Evidence-Based Vocational Rehabilitation Practice.* Presented at the National Council on Rehabilitation Education Conference, San Francisco, CA.

Lui, J., Anderson, C. A., Matthews, P., Nierenhausen, E., & Schlegelmilch, A. (2014). Knowledge translation strategies to improve the resources for rehabilitation counselors to employ best practices in the delivery of vocational rehabilitation services. *Journal of Vocational Rehabilitation, 41*, 137–145.

Murray, C. E. (2009). Diffusion of innovation theory: A bridge for the research-practice gap in counseling. *Journal of Counseling and Development, 87*, 108–116.

Nathan, P. E., & Gorman, J. M. (1998). *A guide to treatments that work.* New York, NY: Oxford University Press.

Odom, S. L., Brantlinger, E., Gersten, R., Horner, R. H., Thompson, B., & Harris, K. R. (2005). Research in special education: Scientific methods and evidence-based practices. *Exceptional Children, 71*, 137–148.

Parkes, J., Hyde, C., Deeks, J., & Milne, R. (2001). Teaching critical appraisal skills in healthcare settings. *Cochrane Database Systems Review.* Available from https://doi.org/10.1002/14651858.CD001270.

Peterson-Karlan, G. R., & Parette, H. P. (2007). Evidence-based practice and the consideration of assistive technology: Effectiveness and outcomes. *Assistive Technology Outcomes and Benefits, 4*(1), 130–139.

Puddy, R. W., & Wilkins, N. (2011). *Understanding evidence part 1: Best available research evidence. A guide to the continuum of evidence of effectiveness.* Atlanta, GA: Centers for Disease Control and Prevention.

Sackett, D. L., Rosenberg, W. M. C., Gray, J. A. M., Haynes, R. B., & Richardson, W. S. (1996). Evidence-based medicine: What it is and what it isn't. *British Medical Journal, 312*, 71–72.

Schlosser, R. W. (2006). The role of systematic reviews in evidence-based practice, research, and development. *Focus, 15*, 1–4.

Sudsawad, P. (2007). *Knowledge translation: Introduction to models, strategies, and measures.* Austin, TX: Southwest Educational Development Laboratory, National Center for the Dissemination of Disability Research.

Tannenbaum, S. (2005). Evidence-based practice in mental health: Practical weaknesses meet political strengths. *Journal of Evaluation in Clinical Practice, 9*, 287–301.

Tansey, T., Bezyak, J., Chan, F., Leahy, M., & Lui, J. (2014). Social-cognitive predictors of readiness to use evidence-based practice: A survey of state vocational rehabilitation counselors. *Journal of Vocational Rehabilitation, 41*, 127–136.

The Campbell Collaboration, 2018. ⟨www.campbellcollaboration.org⟩.

The Cochrane Collaboration, 2018. ⟨www.cochrane.org⟩.

Thomas, R., Barker, L., Rubin, G., & Dahlmann-Noor, A. (2015). Assistive technology for children and young people with low vision. *Cochrane Database of Systematic Reviews,* (6). Art. No.: CD011350. ⟨https://doi.org/10.1002/14651858.CD011350.pub2⟩.

Tucker, J. A., & Reed, G. M. (2008). Evidentiary pluralism as a strategy for research and evidence-based practice in rehabilitation psychology. *Rehabilitation Psychology, 53*(3), 279–293.

Twyman, J. S., & Sota, M. (2008). Identifying research-based practices for response to intervention: Scientifically based instruction. *Journal of Evidence-Based Practices for Schools, 9*, 86–101.

Van der Roest, H. G., Wenborn, J., Pastink, C., Dröes, R. M., Orrell, M. (2017). Assistive technology for memory support in dementia. Cochrane Database of Systematic Reviews, (6). Art. No.: CD009627. ⟨https://doi.org/10.1002/14651858.CD009627.pub2⟩.

Walker, B. B., Seay, S. J., Solomon, A. C., & Spring, B. (2006). Treating chronic migraine headache: An evidence-based practice approach. *Journal of Clinical Psychology, 62,* 1367–1378.

Wenger, E. (1998). *Communities of practice: Learning, meaning, and identity.* Cambridge: Cambridge University Press.

Wenger, E., McDermott, R., & Snyder, W. M. (2002). *Cultivating communities of practice: A guide to managing knowledge.* Boston, MA: Harvard Business School Press.

Winch, S., Henderson, A., & Creedy, D. (2005). Read, think, do!: A method for fitting research evidence into practice. *Journal of Advanced Nursing Practice, 50,* 20–26.

Yousefi-Nooraie, R., Dobbins, M., & Marin, A. (2014). Social and organizational factors affecting implementation of evidence-informed practice in a public health department in Ontario: A network modeling approach. *Implementation Science, 9*(29). Available from https://doi.org/10.1186/1748-5908-9-29.

Consideration of underserved groups and lower resourced environments

Catherine A. Anderson

Rehabilitation Research and Training Center on Evidence-Based Practice, University of Wisconsin-Madison, Madison, WI, United States

CONSIDERATION OF UNDERSERVED GROUPS

Individuals with disabilities represent considerable diversity in race, ethnicity, gender, and socioeconomic status. Many rehabilitation and healthcare service delivery systems now emphasize the need to identify historically unserved and underserved populations and emphasize better engagement and service to these individuals across a variety of need areas (Anderson, Owens, & Nerlich, 2017). The need to demonstrate efficacy of specific approaches is growing and aligns with ensuring equity and fidelity in practice, while allowing enough flexibility for professionals to appropriately address individualized needs.

Why is it important to take the needs of underserved groups into consideration? There is growing acknowledgment that underserved populations are disproportionately represented among low-income populations with limited access to resources. Clinicians and practitioners play a critical role in appropriately assessing and referring for assistive technology (AT) services designed to support individuals with disabilities achieve and maintain their work, social, leisure, and independent living goals. Without access to accommodations such as AT, the likelihood of successful competitive integrated employment outcomes decreases, thereby further limiting the likelihood of achieving financial empowerment. While employment alone does not necessarily move individuals toward financial empowerment, it is a critical component that must be considered in the majority of cases.

DEFINING UNDERSERVED POPULATIONS

It is helpful to understand the contemporary conceptualization of health when discussing the terms unserved and underserved. The definition of health has evolved over the years but the most recent description used by the World Health Organization is "Health is a state of complete physical, mental and social well-being,

Assistive Technology Service Delivery. DOI: https://doi.org/10.1016/B978-0-12-812979-1.00014-X

and not merely the absence of disease or infirmity; health is an individual right and a social justice issue; health is a public good; and governments have a responsibility for the health of their peoples" (World Health Organization, 2010).

Health inequities are the unjust differences in health between persons of different social groups and can be linked to forms of disadvantage such as poverty, discrimination, and lack of access to services or goods (WHO, 2010). Numerous equity indicators are used to distinguish groups and individuals including socio-economic status, education, place of residence (rural, urban, etc.), race or ethnicity, occupation, and gender. Equity indicators frequently overlap and isolating the primary cause can be challenging given that the causes frequently overlap. A good example of this is access to quality education. Education increases literacy, which involves the ability to obtain, read, understand, and use information. Literacy results in access to higher wage jobs and professional opportunities, subsequently resulting in higher income levels. Consequently, lack of opportunity to receive a quality education may result in lower literacy with compounding results (WHO, 2010). Additional examples of underserved populations include those living in rural or remote areas, culturally diverse populations, people with multiple disabilities, adjudicated youth and adults, youth in Foster Care, and individuals experiencing poverty with a specific emphasis on generational poverty.

POVERTY AND DISABILITY

Poverty is a term than can be used interchangeably with underresourced in many situations. It is important to recognize that individuals with disabilities are disproportionately represented among those living in poverty with adults between the ages of 21 and 64 years being twice as likely to live at or below the federal poverty level (28.2%) as those without disabilities. Almost half (47%) of adults whose income poverty lasts for at least 12 months, and nearly two-thirds (65%) of people who live in poverty long term, have one or more disabilities (Erickson, Lee, & von Schrader, 2017). In order to effectively serve those experiencing poverty, rehabilitation professionals must develop a cultural awareness and competencies around in serving this population (Anderson et al., 2017).

Poverty and disability interact and increase the impact of issues such as food insecurity, difficulty paying bills, lack of reliable transportation, and inadequate housing and medical care. These factors also make it harder to access education, find and hold a job, be independent, and fully participate in society. Poverty can contribute to poor physical or mental health, which in turn makes finding a job an even greater challenge—a cycle that can persist across generations. The United States Centers for Disease Control's Healthy People initiative has developed a 10-year framework for tracking and reporting on the nation's health goals and objectives as part of a broader focus on identifying and addressing social determinants of health (National Center for Health Statistics, 2016). The overarching goals of the Healthy People 2020 framework include helping people to

1. Attain high-quality, longer lives free of preventable disease, disability, injury, and premature death;
2. Achieve health equity, eliminate disparities, and improve the health of all groups;
3. Create social and physical environments that promote good health for all; and
4. Promote quality of life, healthy development, and health behaviors across all life stages.

POVERTY AS A CONTRIBUTING FACTOR TO CHRONIC ILLNESS AND DISABILITY

Individuals with disabilities can certainly remain healthy in the context of disability as noted in the World Health Organization's contemporary definition of health cited earlier in this chapter. However, a lack of resource can increase the risk factors associated with poorer health and subsequently increase the risk of acquiring or developing disability. Although poverty can place an individual at greater risk for illness and disability as well as intensify the effects of disability, little focus has been placed on how poverty contributes to both physical and psychiatric disability and subsequently how the burdens of poverty and disability perpetuate patterns of inequality. Researchers have conjectured that people with disabilities are likely to experience additional health-related expenditures that negatively affect their economic well-being as well as experience insufficient food and nutrition, poor housing conditions, and inadequate access to health care throughout childhood which may exacerbate existing health conditions such as asthma, obesity, attention deficit hyperactivity disorder, and other physical and mental health issues resulting in disability in adulthood (Livermore & Hill, 2002; Lustig & Strauser, 2007; Olin & Dougherty, 2006; She & Livermore, 2007). Although many low-income individuals may not identify as having a disability, the likelihood of poor health and healthcare access resulting in chronic illness and/or disability is greatly increased in this population.

SOCIAL DETERMINANTS OF HEALTH

Western societies define and classify disability in terms of physical, cognitive, sensory, or psychological impairment. Those experiencing further intersections of race, poverty, gender, age, and victimization are increasingly vulnerable to a long-term future of economic insecurity.

People's lives include many factors beyond their direct control but present significant influence on health. Most notably, income and social status are linked to better health; with gaps in wealth subsequently linked to gaps in health. Low education levels are also associated with poorer health and increased stress, with ongoing stress serving as a catalyst for subsequent physical and mental health concerns. The physical environment involving access to clean air and water, healthy workplaces, and safe homes and communities are all associated with good health. Social support networks including family, friends, communities, culture, customs, and traditions also influence health. Additionally, people who are employed tend to be healthier, particularly those who have access to safe and healthy work conditions and access to health services that prevent and treat disease while enhancing full participation in work, life, and community are equally important (WHO, 2010).

Youth with disabilities are more likely to live in low-income single parent families and families of racial minority backgrounds thus facing the compounding educational employment and social challenges associated with poverty (Hughes & Avoke, 2010; Parish, Rose, Grinstein-Weiss, Richman, & Andrews, 2008). Youth with disabilities as well as those with significant disabilities are more likely to be placed in restrictive educational settings and are less likely to receive access to appropriate services (Artiles, Kozleski, Trest, Osher, & Ortiz, 2010). As rehabilitation professionals, it is imperative that we are aware of these cultural and systemic biases, and the underlying social determinants of health, as the foundation of our field is to enhance access and full inclusion.

CULTURAL COMPETENCE

As cultural diversity grows globally, so culturally competent knowledge and skills become increasingly important for healthcare and rehabilitation professionals (Preposi Cruz, Estacio, Bagtang, & Colet, 2016; Sairanen et al., 2013). Cultural competence among professionals is foundational in reducing disparities through culturally sensitive and unbiased care, contributing to consumer satisfaction and improved outcomes (American Association of Colleges of Nursing, 2008; Shen, 2015). Culturally competent care is defined as "the care that is responsive and reactive to the diversity of the patient population and cultural factors that can affect health and health care, such as languages, communication styles, beliefs, attitudes, and behaviors" (Murphy, 2011).

While various frameworks for better understanding cultural competence have been introduced over the years, they share commonalities in recognizing (1) an awareness of human dignity, (2) an ability to provide nonjudgment and nondiscrimination care for all, and (3) recognizing that development of cultural competence is a lifelong process (Repo, Vahlberg, Salminen, Papadopoulos, & Leino-Kilpi, 2016; Sagar, 2012; Shen, 2015). Campinha-Bacote (2002, 2007) developed the *Process of Cultural Competence in the Delivery of Healthcare Services* model that delineates cultural competence as an ongoing process involving relationships among a number of key constructs critical in providing service to individuals from various cultures:

Cultural awareness—ability to recognize one's own cultural and professional background regarding personal beliefs, values, and perceptions.

Cultural knowledge—the process of gaining deeper understanding of others' cultures, worldview, and disability.

Cultural skill—the ability to collect relevant cultural information regarding the individual's presenting issue and performing culturally sensitive assessments, modifying as needed to respect cultural differences.

Cultural encounters—the process in which the professional engages in cross-cultural interactions with individuals from culturally diverse backgrounds, thus increasing the opportunity for service providers to develop an appreciation for varying cultures through first-hand information and experience.

Cultural desire—a motivation by rehabilitation and healthcare professionals to engage in the aforementioned constructs and practice in a culturally appropriate manner in every aspect of care.

PREDICTORS OF CULTURAL COMPETENCE

The strongest predictors of cultural competence among nursing students include (1) prior diversity training, (2) living in an environment with people of diverse racial and ethnic backgrounds, and (3) experiencing caring for culturally diverse populations (Preposi Cruz et al., 2016).

Integrating cultural competence elements into rehabilitation professionals' education is important and recommended. Given that AT specialists come to the profession from a variety of diverse educational and experiential backgrounds, it poses challenges in systematically integrating it into programs. Alternatively, cultural competence can be integrated into ongoing professional development activities, which allows for broader reach and scope in getting information to the individuals currently in practice. Assessment of cultural competence development is recommended as a means through which to measure when and how knowledge is acquired, applied, and shared. Sairanen et al. (2013) recommend that the following competency areas be integrated into rehabilitation professionals' educational curriculum and ongoing professional training:

- Demonstrate cultural awareness and the ability to apply it in practice
- Demonstrate knowledge of ethnic differences which can affect the assessment of client health status and the subsequent treatment and care of individuals
- Assess the needs and health status of individuals taking into account cultural beliefs and practices
- Engage confidently in caring for people whose culture is different to their own

Furthermore, they recommend framing the competency areas across three domains of (1) cultural awareness, (2) cultural knowledge, and (3) cultural skills to help simplify the exchange and acquisition of knowledge for learning and assessment purposes (Sairanen et al., 2013).

Cultural awareness—includes terminology; political, social, and cultural factors; self-awareness, communication techniques, layers of culture, cultural conflicts, human rights, ethical foundation, and culture shock.

Cultural knowledge—includes health literacy; health and illness beliefs in caring and curing; assessment methods; cultural theories and models; understanding of contemporary migration and immigration issues; and lifespan events.

Cultural skills—includes skills to deliver culturally sensitive services; provide culturally safe interactions; and evaluate and assess interventions for their level of cultural appropriateness.

This then leads into the next question—does participation in cultural competence professional development work? Chipps, Simpson, and Brysiewicz (2008) conducted a systematic review and found positive changes in cultural knowledge gain and attitude to demonstrate the most frequently noted change. Importantly, consumer satisfaction improved after rehabilitation professionals participated in training of this nature (Fig. 14.1).

Assessing cultural competence. The *Cultural Competence Assessment Instrument (CCAI-UIC)* developed by Suarez-Balcazar et al. (2011) and recommended for use by rehabilitation professionals consists of 24 items, grouped into three primary domains: (1) awareness, (2) organizational support for multicultural practice, and (3) skills. You might notice that cultural competence awareness and skills have been mentioned previously in this chapter. The addition of organizational support to practice in a culturally competent manner is unique to this tool and noted as an important component to consider. Well-intentioned professionals may be committed to ongoing learning and development of increasingly culturally competent practice, but if this is not supported within the context of the organizational culture where they are employed, it can be very challenging to achieve.

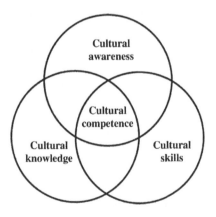

FIGURE 14.1

The three domains of cultural competence.

CONCLUSION

In conclusion, it is becoming increasingly important for rehabilitation professionals to better understand the nuances of cultural competency and proactively engage in professional training to continually develop skills and knowledge in this area. Furthermore, a commitment to ongoing development of cultural competence provides a strong foundation for creative thinking in engaging and serving individuals living in underresourced areas and/or those experiencing poverty.

REFERENCES

American Association of Colleges of Nursing. *Cultural competency in baccalaureate nursing education*. (2008). Retrieved from ⟨http://www.aacn.nche.edu/Education/pdf/competency.pdf⟩.

Anderson, C. A., Owens, L., & Nerlich, A. (2017). Poverty, disability and vocational assessment of youth with disabilities. *Vocational Evaluation and Work Adjustment Association Journal*, *41*(2), 3–13.

Artiles, A. J., Kozleski, E. B., Trest, S. C., Osher, D., & Ortiz, A. (2010). Justifying and explaining disproportionality, 1968–2008: A critique of underlying views of culture. *Exceptional Children*, *76*, 279–299.

Camphina-Bacote, J. (2007). *The process of cultural competence in the delivery of healthcare services: The journey continues* (5th ed). Cincinnati, OH: Transcultural C.A.R.E. Associates.

Campinha-Bacote, J. (2002). The process of cultural competence in the delivery of healthcare services: A model of care. *Journal of Transcultural Nursing*, *13*(3), 181–184.

Chipps, J. A., Simpson, B., & Brysiewicz, P. (2008). *The effectiveness of cultural-competence training for health professionals in community-based rehabilitation: A systematic review of literature*. Worldviews on Evidence-Based Nursing, Second Quarter.

Erickson, W., Lee, C., & von Schrader, S. (2017). *Disability statistics from the American Community Survey (ACS)*. Ithaca, NY: Cornell University Yang-Tan Institute (YTI). Retrieved from Cornell University Disability Statistics website: www.disabilitystatistics.org.

Hughes, C., & Avoke, S. (2010). The elephant in the room: Poverty, disability, and employment. *Research & Practice for Persons with Severe Disabilities*, *35*(1–2), 5–14.

Livermore, G., & Hill, S. (2002). *Changes in health care expenditures and the employment of people with disabilities, 1987–1997*. Washington, DC: Cornell University Institute for Policy Research, Paper prepared for the National Institute for Disability and Rehabilitation Research.

Lustig, D. C., & Strauser, D. R. (2007). Causal relationships between poverty and disability. *Rehabilitation Counseling Bulletin*, *50*(4), 194–202.

Murphy, K. (2011). The importance of cultural competence. *Nursing Made Incredibly Easy*, *9*(2), 5.

National Center for Health Statistics. *Healthy people 2020 progress review: Improving health outcomes through inclusion and participation*. (2016). Retrieved from

⟨https://search.cdc.gov/search/?subset = NCHS&query = Health + People + 2020&utf8 = % E2%9C%93&affiliate = cdc-main&sitelimit = www.cdc.gov%2Fnchs%2F⟩.

Olin, G., & Dougherty, D. (2006). *Characteristics and medical expenses of adults 18 to 64 years old with functional limitations, combined years 1997−2002*. Rockville, MD: U.S. Department of Health and Human Services, Agency for Healthcare Research and Quality, Agency for Healthcare Research and Quality Working Paper 06002.

Parish, S. L., Rose, R. A., Grinstein-Weiss, M., Richman, E. L., & Andrews, M. E. (2008). Material hardship in U.S. families raining children with disabilities. *Exceptional Children*, *75*, 71−92.

Preposi Cruz, J., Estacio, J. C., Bagtang, C. E., & Colet, P. C. (2016). Predictors of cultural competence among nursing students in the Philippines: A cross-sectional study. *Nurse Education Today*, *46*, 121−126.

Repo, H., Vahlberg, T., Salminen, L. K., Papadopoulos, I., & Leino-Kilpi, H. (2016). The cultural competence of graduating nursing students. *Journal of Transcultural Nursing*. Available from https://doi.org/10.1155.2013/929764.

Sagar, P. L. (2012). *Transcultural nursing theory and models: Application in nursing education, practice, and administration*. New York, NY: Springer.

Sairanen, R., Richardson, E., Kelly, H., Bergknut, E., Koskinen, L., Lundberg, P., ... De Vlieger, L. (2013). Putting culture in the curriculum: A European project. *Nurse Education in Practice*, *13*, 118−124.

She, P., & Livermore, G. (2007). Material hardship, poverty, and disability among working-age adults. *Social Science Quarterly*, *88*(4), 970−989.

Shen, Z. (2015). Cultural competence models and cultural competence assessment instruments in nursing: A literature review. *Journal of Transcultural Nursing*, *26*(3), 308−321.

Suarez-Balcazar, Y., Rodakowsk, J., Balcazar, F., Garcia-Ramirez, M., Taylor-Ritzler, T., Willis, C., & Portillo, N. (2011). Development and validation of the cultural competence assessment instrument: A factorial analysis. *Journal of Rehabilitation*, *77*(1), 4−13.

World Health Organization. (2010). Consultation on implementing action on social determinants of health to reduce health inequities: The contribution of collaborative work between sectors. In: *Meeting Report, September 2010*.

On technology abandonment or discontinuance

15

Ray Grott

Rehabilitation Engineering and Assistive Technology (RET) Project, San Francisco State University (SFSU), San Francisco, CA, United States

INTRODUCTION

Some researchers have suggested that end-users of product design are nothing but a predictable bundle of reflexes and impulses that can be torqued, tuned, and tweaked in order to do the bidding—and the buying—prescribed by a consumer savvy cabal of designers, engineers, and marketers

Almquist and Lupto (2010, p. 3).

Given the rates of assistive technology (AT) which go unused following implementation, AT professionals would have to disagree with the above statement. Sometimes the stock or customized AT solution just doesn't work as desired. Following a solid problem-solving and design process, including good communication and collaboration with the employer and employee, and drawing on expert fabrication skills, doesn't provide immunity from disappointment. This disuse of AT is common enough to have its own term "technology abandonment," though some have suggested that "discontinuance" would be a better description (Lauer, Longenecker Rust, & Smith, 2018). There are many possible reasons for this (Scherer, 2005). We will discuss some of them in this chapter.

REASONS FOR NOT USING THE PROVIDED TECHNOLOGY
SOLUTION DIDN'T WORK AS DESIRED

In this situation, one must go back and explore the details of the situation, exploring what might have been missed in the problem analysis or design process. (Assuming, of course, that some form of a problem-solving methodology described in Chapter 9: Problem Solving was used.) Maybe a key measurement was wrong or based on a guess. Maybe there wasn't sufficient time to build and test a prototype or mockup or get a loaner device to try. Maybe a simple tweak will fix the problem, or maybe it's back to the drawing board.

Assistive Technology Service Delivery. DOI: https://doi.org/10.1016/B978-0-12-812979-1.00015-1

POOR COLLABORATIVE PROCESS

A lot has been written about the abandonment or discontinuance of technology use. Almost all that I have read list as one of the culprits the end user not being appropriately involved in the process of identifying or selecting the AT device; not using a person-centered approach. If this is not done adequately, there can be a mismatch between the solution and the user's needs and expectations, increasing the odds of a poor outcome.

POOR DESIGN

Obviously, a device or modification that doesn't do what it promises to do is poorly designed. But there is another element to a workable design. It is common for people to become so familiar with our project and knowledge in any arena that we forget the audience when we present ideas, information, or instructions and engage in design work. What seems obvious to us because we are so intimately involved with the material may come across to others as complicated and difficult to understand. If a device, process, AT solution, or other accommodation seems too complicated, too cumbersome, or demands too much of a learning curve or cognitive load, it is more likely to be rejected.

LACK OF SUPPORT

Lack of support can take many forms. There can be delays or mistakes in the acquisition of needed items that build frustration and tension. Incomplete or inadequate provision of the information technology or other support services required to implement the accommodation can lead to imperfect results that are blamed on the AT device or system. Management can fail to encourage and provide training and ramp-up time for the employee to use the AT successfully. Supervisors can fail to promote an understanding and supportive attitude among fellow workers if their involvement is needed in making the accommodation work.

INADEQUATE TRAINING

This is most common with software and electronic tools but can also apply when more complicated physical solutions are implemented. Too often the employer in a busy company agrees to a solution but is reluctant to allow adequate time for the busy employee to train and practice and gradually come up to speed. Sometimes, the employer is flexible, but the employee feels rushed and anxious about completing their work tasks and will find reasons to avoid being available to schedule training or practice. In other situations, the evaluator is asked to implement the solution but does not have the needed knowledge or training skills.

TASKS, GOALS, PROCEDURES, OR TOOLS CHANGED

There can be many components to a job, and a change in any one can cause problems. For example, one company got a new, larger centralized networked printer and their wheelchair-riding employee could no longer access it to scan in documents that she was unable to manipulate by hand. (In this instance, providing her with a personal scanner in her cubicle solved the problem.) Another person was doing fine using speech recognition software instead of typing until the company changed to a new database system that didn't work well with that input method, presenting a more complicated challenge. Revising the way that paychecks were printed and processed made a custom folding jig for a bookkeeper with limited hand useless. Alternately, changes in tools and methods make a job easier for someone with a disability. For instance, shifting to an electronic medical record system helped a clerk who lacked the arm strength needed to pull files but who had good typing skills transition to working with these records.

SKILLS OR ABILITIES ARE INADEQUATE OR CHANGE

As mentioned earlier, disability, chronic illness, or serious injury is not a constant. Sometimes an employee's physical or cognitive abilities improve, or get worse. Building in flexibility is not always possible or sufficient to accommodate the change. Sometimes as well, it appears that the employee was not prepared to accept the technology-related changes. I once implemented a set of accommodations for someone who had a stroke and only later discovered problems she had with learning new multistep processes that were not apparent in evaluation or preliminary off-site training.

SOLUTION REJECTED FOR SUBJECTIVE REASONS

This can be the most frustrating outcome. There are times when I've provided what I thought was a perfectly appropriate and functional customization and the employee decided they did not like it. It wasn't because it didn't work or meet the criteria. They just didn't like it. Sometimes they were just more comfortable doing things the old way. Sometimes they felt it made their disability and need for accommodation more visible to their coworkers. One then has to ask whether there were some unexplored questions in the problem-solving process related to personal preferences, or work settings and interactions with other employees. Other overlooked elements can include cultural beliefs, feelings of shame, or a fear of stigma.

There may be other issues playing out behind the scenes. Examples of this can be a desire to see the solution fail because of anger with the supervisor or company, or an unstated goal of pushing for a different job as an accommodation or qualifying for disability retirement. Or vice versa—the supervisor may feel that the solution is inadequate because they want to get the employee out of their department.

AN EXAMPLE OF SUBJECTIVE REASONS FOR TECHNOLOGY NOT BEING USED

A woman working in a tissue culture lab at a hospital developed significant repetitive strain injuries in her hands from the many tasks that required precise gripping and tool manipulation in awkward positions. The workers compensation insurance company brought me in to review the situation and this author ended up making a number of custom devices such as a scalpel holder that allowed her to use a full hand grip in a thumbs-up neutral position. I also made a rig to enable her to control a syringe using her whole hand and arm to raise and lower a lever arm (as with a pump handle). The employee was involved in testing and refining the prototypes, and everything worked as desired. However, it turned out that the employee's personal goal was to work part time or go out on unemployment and get retraining for a different line of work, and the employer was not willing to allow part-time work and was hoping she would resign. As the consultant, I just had to do my best with what I was requested to do and walk away from the drama. My hope is that these efforts at least demonstrated to the employer and insurance carrier that such accommodations are possible.

THE USER CAME UP WITH AN ALTERNATE SOLUTION

Not all lack of use can be considered a failure of the original solution. As mentioned, sometimes the need changes or no longer exists. In other instances, the employee might have devised or discovered a more suitable method of performing the task. Or found ways to sidestep the task altogether.

My favorite example is that of a woman wheelchair rider with limited use of her arms and hands. She had a job restocking DVDs in a video rental store (before Internet access forced most of these to close). Her challenge was not being able to pick up cases that she or others dropped on the floor. I devised a simple device based on the concept of adding a suction cup to the end of a stick. I was very proud of my creative solution and was disappointed when I met with her a couple months later and she told me she was no longer using it. When I asked her why, she explained that she had become friendly enough with the rest of the staff that she just asked for assistance when needed. I decided that if using the device enabled her stay on the job long enough to reach a point of social integration with the staff, it was a successful outcome, even if it was no longer being used. (Clearly, if picking up these items from the floor was a significant portion of her job, asking others for assistance would not have been an appropriate solution unless approved by her manager.)

There are several reasons that AT users decide to accept or reject their assistive devices. Results showed that 29.3% of all devices were completely abandoned. Mobility aids were more frequently abandoned than other categories of devices, and abandonment rates were highest during the first year and after 5 years of use. Four factors were significantly related to abandonment:

- Lack of consideration of user opinion in selection;
- Ease of device procurement;
- Poor device performance;
- Change in user needs or priorities (Phillips & Zhao, 1993).

To these four reasons, De Barros, Duarte, and Cruz would add another factor:

- Stigma (based on how AT is presented and perceived) (De Barros, Duarte, & Cruz, 2011).

Desmet and Hekkert discuss user−product interaction as being comprised of three primary elements—the experience of meaning and the aesthetic experience both of which inform the user's affective response to a product. They also offer a basic model of appraisal which gives rise to emotional responses—the product and a user's concerns inform product appraisal which in turn informs and elicits an emotional response. How we perceive a product and the meaning we ascribe to our appraisal influence how we feel about the product (Desmet & Hekkert, 2007). How we perceive a product makes a difference in how we respond to it.

De Barros, Duarte, and Cruz conducted a study wherein they presented each of two groups a magazine featuring assistive devices, one group received the magazine designed as being marketed to disabled and elderly and the other designed in a standard product marketing manner. The authors found that item utility was readily accepted by both groups but the assistive or supportive nature of the labeling of products negatively impacted their perception of the products (De Barros et al., 2011).

CONCLUSION

Designers need to recognize that consumers are central to the design process and cannot be reduced to mere consumer product automatons. The user experience is influenced by many variables not the least of which are social affirmation and perception of AT; technology-related policies and services need to emphasize consumer involvement, long-term needs (Phillips & Zhao, 1993), and the elimination of stigmatizing labels and AT presentations opting instead for a more general product marketing strategy. This would serve to reduce device abandonment and enhance consumer satisfaction.

REFERENCES

Almquist, J., & Lupto, J. (2010). Affording meaning: Design-oriented research from the humanities and social sciences. *Design Issues, 26*(1), 3−14.

De Barros, A., Duarte, C., & Cruz, J. (2011). The influence of context on product judgement—Presenting assistive products as consumer goods. *International Journal of Design, 5*(3), 99−112.

Desmet, P., & Hekkert, P. (2007). Framework of product experience. *International Journal of Design, 1*(1), 57−66.

Lauer, A., Longenecker Rust, K., & Smith, R. O. (2018). *ATOMS project technical report—Factors in assistive technology device abandonment: Replacing "abandonment" with "discontinuance"*. University of Wisconsin-Milwaukee. Retrieved from www.r2d2.uwm.edu/atoms/archive/technicalreports/tr-discontinuance.html.

Phillips, B., & Zhao, H. (1993). Predictors of assistive technology abandonment. *Assistive Technology, 5*(1), 36.

Scherer, M. J. (2005). *Living in the State of Stuck: How Assistive Technology Impacts the Lives of People with Disabilities, Fourth Edition*. Cambridge, MA: Brookline Books. (First Edition published in 1993; Second in 1996; Third in 2000).

Assistive technology techniques, tools, and tips

16

Anthony Shay[1] and Marcia Scherer[2]

[1]*Capacity Building Specialist, Assistive Technologist, and Rehabilitation Specialist, University of Wisconsin-Stout Vocational Rehabilitation Institute (SVRI), Menomonie, WI, United States*
[2]*Institute for Matching Person and Technology, University of Rochester Medical Center; Physical Medicine and Rehabilitation and Senior Research Associate, International Center for Hearing and Speech Research, Webster, NY, United States*

ASSISTIVE TECHNOLOGY TOOLBOX

We open our toolbox to some basic information of interest to professionals working in the disability-employment and assistive technology (AT) fields. Information does not always translate well between these fields. It is our hope that areas of overlap as reflected in these bits and pieces of information will facilitate a better understanding for both. The toolbox offers a snapshot of case management models, perspectives on why standardization matters in service delivery, and case noting in service delivery. We take a broad-brush approach to delineating functional employment elements—the general job expectations placed on us while working—as an overview for disability-employment and AT professionals alike.

CASE MANAGEMENT

The Commission on Rehabilitation Counselor Certification defines case management in terms of the process being a merging of counseling and managerial concepts and skills; derived from intuitive and researched methods; efficient decision making as a foundation for a proactive practice; and the counseling role as focused on coordinated service delivery, interaction with others, monitoring client activity, and problem solving (Commission on Rehabilitation Counselor Certification, 2018, p. 6). This harkens back to 1964 when Albert Thompson stated, "Goals are necessary for motivated behavior but they can also be restrictive ... we have a changing individual and a changing environment and the real decision is how to plan for change" (p. 223). According to Thompson, when we work with people who have disabilities, we must account for the changes in context they will experience. This requires a willingness (maybe an eagerness on the part of professionals) to increase our knowledgebase as a critical factor in maintaining competence toward moving people in the direction of their goals. We work to move them forward even when

Assistive Technology Service Delivery. DOI: https://doi.org/10.1016/B978-0-12-812979-1.00016-3

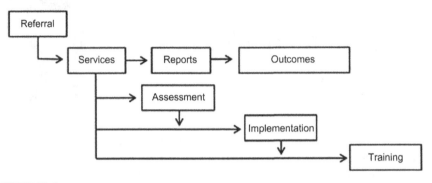

FIGURE 16.1

Birds-eye view of the AT service delivery process. *AT,* assistive technology.

there is a lack of goal striving clarity. We can facilitate the development of clearer goals by introducing individuals to situations and experiences which help them better understand the ramifications of their employment-related choices. Thompson extolls the virtues of this kind of situational decision-making as a precursor to informed choice. The responsibilities we have inherent in case management necessitates the provision of quality information, a comprehension of context-dependent adaptive decision-making skills, and the capacity to address thinking errors in the people we serve. All of these items are crucial to person-centered services and effective case management (1964, p223).

The case management process mirrors the service delivery process (see Fig. 16.1 for a birds-eye view of the AT service delivery process). The AT service delivery process involves case management across nine discrete steps: (1) referral, (2) intake and initial assessment, (3) systematic assessment, (4) plan development, (5) recommendations and report, (6) technology procurement/development, (7) implementation, (8) follow-along, and (9) follow-up/rereferral. Table 16.1 provides an overview of the elements which must be considered at each step of the process (not an exhaustive list) (Shay, Anderson, & Matthews, 2017, p. 79). People with disabilities looking for work tend to find themselves dealing with many service providers and systems on their path to employment. We as providers encounter these frequently as well. These systems likely function utilizing differing case management processes. It is important to consider the underlying contexts, knowledge domains, and the discipline within which service providers operate.

Wong lists several models of case management from the contemporary and mental health domains. The contemporary domain consists of role-based, organization-based, and responsibility-based case management models. The models are defined in terms of the roles and responsibilities characterized by the model.

Role-based: This approach tends to emphasize the individual as the focus of case management activities. An example of this would be a professional employment and rehabilitation counselor in an employment-rehabilitation system.

Organization-based: Attention is placed on the provision of services within organizations such as independent living centers, community action programs, AT

Table 16.1 Case Management Considerations Within the AT Service Delivery Process

AT Service Delivery Process Steps	Case Management Considerations
1. Referral	Collect referral information and contact consumer; basic information gathered on disability/limitations, referral purpose, assign staff
2. Intake and Initial Assessment	Follow-up with and add to information received during referral, determine service delivery needs, begin gathering information regarding functional skills/limitations, employment goal, begin considering AT plan objectives
3. Systematic Assessment	Complete a formal AT assessment including a comprehensive needs assessment, feature matching to include demonstrations, trial, and simulations as necessary
4. Plan Development	Develop intervention strategies and build in timeframes and progress measures toward meeting consumer informed choice and rehabilitation team expectations
5. Recommendations and Report	Write a review of the assessment, AT plan, recommendations, implementation plan and toward approval to move to next steps
6. Technology Procurement and Development	Ordering or purchasing the AT devises, systems, parts, and materials with customization or fabrication as necessary, toward implementation
7. Implementation and Training	Implementing or installing AT as outlined in the implantation plan including training as needed
8. Follow-Along and Case Termination	Over the short-term measure and document outcomes, reassess as necessary and provide necessary services toward closing the case
9. Follow-Up and Re-referral	Over the longer-term following case closure, measure and document outcomes, reassess as necessary (e.g. need for additional or remedial training) including the need for a new referral to the referral/funding or another source

service providers, or community rehabilitation programs (e.g., extended employment centers).

Responsibility-based: This approach centers on the transition from professional activity to that of nonprofessionals (e.g., family, friends, volunteers, and support networks) (Wong, 2009; Woodside & McClam, 2003).

Case management models in the mental health domain differ slightly from those of the contemporary models. Mental health case managements models include standard, rehabilitation-oriented, and intensive case management models (Wong 2009; Mueser, Bond, Drake, & Resnick,1998).

Standard: In the standard approach to case management, the professionals provide advocacy fiscal services, plan, organize, and implement services for those they serve.

Rehabilitation-oriented: This model represents a shift away from the medical model to a more positive psychology frame emphasizing an individual's strengths and deemphasizing pathology. A hallmark of this approach is building an effective working alliance.

Intensive: The intensive approach to case management shifts responsibility away from a single professional in favor of a team-based approach to managing a case. This limits case load sizes and provides for greater active involvement of team members with those they serve. This may include home or work-based interventions, medication management, round-the-clock care, and the provision of activities of daily living (Wong, 2009) (Fig. 16.1).

CASE MANAGEMENT DELIVERABLES

Case management facilitates the process of understanding and managing caseloads. Depending on the employment and disability setting, professionals may experience high caseloads. Choosing a case management model that addresses the needs of the organization and the individuals served is key component to achieving successful outcomes. These include the following:

- Empowers people with disabilities and reinforces self-reliance;
- Reduces the return rate of individuals for services;
- Sets a measurable standard for high-quality service delivery for staff;
- Sets an expectation of professional responsibility and accountability across services/service teams;
- Reduces the case service costs (e.g., shorter case duration, reduced case costs, reduced staff time on case);
- Increased clarity regarding roles, responsibilities, and expectations regarding interdisciplinary/interagency team members;
- Ensures the provision of timely and appropriate services;
- Effective case management (and caseload management) can reduce the turnover rate in an organization which can have an impact on all aspects of service delivery outcomes.

Moore finds the integration of case management with service delivery lays the groundwork for effective coordination of programmatic components, both formal and informal. With integration, programs are better able to deal with "resource deficits and growth in the demand for services" (Moore, 1992, p. 422). Standardizing processes is one way to introduce stability into a system.

ON STANDARDIZATION

Process standardization can be defined as the improvement of operational performance, cost reduction through decreased process errors, facilitation of communication, profiting from expert knowledge (Wüllenweber, Beimborn, Weitzel, & König, 2008, p. 212), and providing flexibility without sacrificing organizational

controls (Røhnebæk, 2012). By definition, standardization is a benefit to an organization. However, staff responses to these efforts are not always effective.
Røhnebæk offers three ineffective responses to standardization efforts:

- Pragmatic ignorance: ignoring the fact that current standard operating
  procedures do not adequately address the current work conditions. The focus
  is on work tasks to the exclusion of process. In other words, the nose-to-the-
  grindstone approach;
- Compliance: involves increased stress as staff work to deal with the mismatch
  of the current situation with standard operating procedures. The tendency here
  is to redouble efforts to apply prescribed policy and procedures. This is the
  proverbial square peg pounded into the circular hole phenomenon;
- Adaptation: perceiving that standard operating procedures could be helpful but
  require some minor modifications. Resistance to standardization may lead to
  modifications to a prescribed process which leads to an overly complex
  system which makes managerial oversight problematic. This is the "road to
  perdition is paved with good intentions" approach (Røhnebæk, 2012, pp.
  692–693).

Effective standardization is arrived at through addressing both tasks *and* processes and achieving a balance between them. The tasks staff are engaged in must
reflect the contexts and actual circumstances within which they operate and
engage the people with whom they work. Likewise, organizations develop standard operating procedures as a product of legislative oversight, ethical/best practices, as well as policies and procedures arrived at through evidence-based
practices. Neither tasks nor processes are sacrificed one for the other.

CASE NOTE STANDARDIZATION

Case noting is a critical component of case management in service delivery.
Effective case noting may be an overlooked aspect of case management. The
pressures placed on service delivery systems can be manifold. The structure and
organization of frontline professionals' work may demand they function in a
highly individualized manner. They need flexibility and discretion. At the same
time organizations need efficiency and service delivery based on established rules
and principles. Often, management finds themselves in the difficult position of
exercising policy and procedures remotely—seeking balance between individuali-
zation and standardization. This balance is achievable. Røhnebæk advocates for
information systems as both offering flexibility and control as a means to "standardized flexibility" (Røhnebæk, 2012, p. 679–680). A discussion of the efficacy
of information systems is beyond the scope of this text. Our discussion will focus
on a general framework for case note standardization.

Case noting provides a formal record which captures the essence of all case-
related activity. It documents stakeholder case activity from the perspective of the
professional maintaining the record. It also allows for a comparison of case note

activity across staff members using the same established policies and procedures while ensuring a comprehensive record of activity. This also reduces susceptibility to errors. Case notes tell the rehabilitation story. They follow people throughout the service delivery process. The narrative reflects stakeholder activity at each stage in the process and written with just enough objective detail to allow for case services continuity and clarity for others who may need to access or manage the case. See Table 6.1 in Chapter 6: Overview of the Service Delivery Process for an example of the steps in both the employment of people with disabilities and the AT processes.

The functions of case noting

Effective communication is necessary for positive outcomes (Meeks, 2001). Case noting, as a case management tool, is an essential component of effective communication in service delivery systems. It is an intentional process whereby meaningful information and consumer contact is recorded. Data within the record is reflective of information regarding, and activities performed by, the subject of the record as well as by relevant stakeholders. Case noting serves many function including (but not limited to)

- Telling the rehabilitation story. Case noting allows us to track an individual as they move through the service delivery process. An effective narrative summary provides adequate detail to allow anyone in the organization to review the casefile and understand what has happened, next steps, and the shared goals and expectations for moving forward;
- Provides a formal record of the organization's service delivery process as it relates to an individual's case;
- Comprehensive case management (e.g., time management/continuity of services, coordination and timeliness of services, staffing assignments);
- Provides management a means to engage in monitoring, quality assurance, outcomes measurement, staff performance, and training activities;
- Provides a means for governmental/legislative and administrative oversight.

The practice of case noting

The practice of case noting, as stated previously, tells the rehabilitation story—the essence and flow of what has transpired in a given case. As a case management tool, it facilitates

- Guidance and counseling;
- Goal/vocational identity striving activity;
- Information and data processing throughout the service delivery process;
- Problem-solving and decision-making;
- Planning, organizing, prioritizing, and monitoring activities;
- Tracking roles and responsibilities (e.g., consumer, case facilitator, rehabilitation team);

- Benchmarking and tracking consumer case progress throughout the service delivery process.

Ethical considerations

Professionals who work with people with disabilities (PWD) seeking employment often work for and/or are members of professional organizations such as the Rehabilitation Engineering and Assistive Technology Society of North America (RESNA) and the Commission on Rehabilitation Counselor Certification (CRCC) which require strict adherence to ethical codes and standards of practice as a condition of employment, licensure/certification, or membership. The language for these codes may vary but the overarching mandate is that code adherents "hold paramount the welfare" (Rehabilitation Engineering and Assistive Technology Society of North America, 2017, p. 15) and accept the "responsibility to provide caring service to individuals with disabilities" (Commission on Rehabilitation Counselor Certification, 2017). In all aspects of service delivery, protecting the privacy and confidentiality of PWD is a primary concern.

CRCC offers several ethical considerations related to case noting (this is not an exhaustive list):

- Timeliness—documentation must occur within a specified timeframe following case activity;
- Adequacy—documentation is completed with sufficient detail;
- Maintenance of documentation accuracy—errors are corrected as soon as they are discovered—either through correcting an existing case note or creating another which clarifies and corrects previous information;
- Documentation is maintained in a secure location ensuring privacy and confidentiality;
- The subject of the record has access to documentation as allowed per policy and law;
- Documentation may be released to third parties following discussion and signed releases by the subject of the record or their legal representative;
- Records disposal occurs per policy and law in a manner that protects privacy and confidentiality;
- Reasonable precautions are observed following case termination, transfer, disasters, or other unforeseen interruptions in services to protect privacy and confidentiality.

A basic model

Using a model can be helpful when case noting. Organization and flow are enhanced. Case notes are easier to write, read and understand saving time for both the writer and the reader. The following points provide a rudimentary framework for the generation of comprehensive case notes. It is comprised of eight essential questions which are to be considered whenever documenting case activity. The expectations of each person and/or agency should be clearly defined. The framework consists of reflecting on and writing about case activity based on the following five Ws and three Hs:

1. **Who** is the person or are the people being referred to or are responsible for or taking on specific roles related to the case activity?
2. **What** exactly is the case activity being documented—what are the next steps to be undertaken?
3. **When** is the specific timeframe for the activity? When did or will it occur?
4. **Where** will the specific case activity occur? Include multiple contexts and transit between contexts.
5. **Why** is the case activity taking place? Provide the rationale behind the need for the activity.
6. **How** will the specific case activity take place?
7. **How** much will the specific case activity cost? Is it reasonable, necessary, and appropriate for the activity?
8. **How** long will the case activity take? What is the total expected duration of the activity? (J. Rogers, Personal Communication, September 2009).

Writing style is also an important consideration. Items such as writing to document important and relevant case information (rather than everything and anything that happens), in the first rather than third person (*I* did ... versus the *[your title here]* did ...), the attitude of the case note (i.e., case notes should not reflect emotion such as having an angry tone), use discretion (what you write in your case notes is a reflection not only of you but the organization for whom you work), objectivity (note only those things observed or heard (do not falsify information or record opinions, unsupported speculation, personal feelings, or value judgements), be brief (be concise and to the point), use appropriate and plain language (do not use jargon, overly technical terms, offensive or prejudiced language). How you present the information is as important as what you are documenting (Wisconsin Division of Vocational Rehabilitation, 2015).

Assistive technology service delivery is highly individualized. Case noting models provide a means to capture as much relevant detail regarding case activity as possible. When taken together with organization policies, procedures, and ethical guidelines, a systematic and comprehensive approach to service delivery which safeguards consumer privacy and confidentiality is achievable.

ASSISTIVE TECHNOLOGY CASE NOTING

Although case noting may occur throughout the assistive technology (AT) service delivery process and different programs may vary somewhat in how they structure AT service delivery, they do share some common points. These may include:

- Referrals and case assignments;
- Funding approvals;
- Consumer contacts (e.g., referrals, intake, scheduling)

- Assessment activity (e.g., feature matching, demonstrations);
- AT plan development;
- Research activity;
- AT procurement activity (e.g., commercially available items, parts, materials);
- Customizing or fabricating AT;
- Installing equipment;
- Training activity;
- Follow-up and follow-along activity;
- Case termination.

Case noting occurs along these nodes of case activity which highlight the consumer rehabilitation story (Greenwood & Roessler, 2018). Although there may be sundry other notes in the case record (e.g., fiscal, contact/scheduling, progress reports from other service providers), the key nodes reflective of a consumer's rehabilitation story should be known to the case facilitator.

THE FUNCTIONAL ELEMENTS OF EMPLOYMENT

An AT assessment addresses prospective (or current) user work tasks and activities, the work environment(s) within which the AT will be used, and the specific assistive technologies the AT professional and the individual determined would be successful in meeting the user's needs as well as the overall demands on an AT device or system. The assessment process, which begins at referral and runs the course of the AT service delivery process, considers some basic elements of on-task behavior. AT (or other accommodations) can be provided to replace, augment, or supplement these functional elements (see Fig. 16.2). Because these activities either occur on a worksite or within the work context, employer concerns and expectations are a central component of AT consideration.

Accuracy: The ability to perform the elements of a task to the standards set for that task by the employer. Given an individual's functional limitations and extant skills, on-task intermediate goals may need to be adjusted until accuracy and production standards can be achieved or job accommodations are provided to meet their needs.

Anticipation: The ability to identify (and modify if necessary) goals and objectives and to determine needs toward maintaining and completing tasks in a safe manner. Although evaluation/assessment may aid in task anticipation--toward planning and organizing--task anticipation is a more fluid and informal process (thinking-on-your-feet).

Completion: The ability to independently follow through and finish tasks based on task planning and employer expectations. Task completion also refers to cleanup of the work space and storage of materials, tools, etc. used in task engagement.

Evaluation/assessment: An individual's ability to evaluate task progress based on feedback and adjust aspects of task engagement to effectively achieve

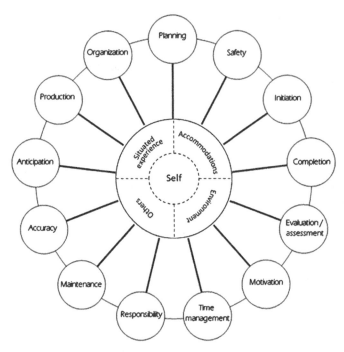

FIGURE 16.2

Accommodating the functional elements of employment.

task goals. Task assessment refers to the ability to assess performance and progress toward goals and to utilize feedback effectively, creative thinking, problem-solving, and rational decision-making toward optimizing task engagement activity. Using an action research model (i.e., planning, acting, observing, reflecting, and repeating as necessary) can facilitate this process.

Initiation: The ability to independently begin task activity, typically following planning and organizing. Task initiation also refers to setup and staging of the work spaces used in the task engagement activity.

Maintenance: Ability to remain engaged in on-task behavior and to meet goals and objectives based on feedback received. Task maintenance also refers to ongoing cleanup, interim storage, staging, and movement to and from task contexts and processes preceding, during, and following the task engagement activity. Essentially, task maintenance involves an individual's efforts toward workflow continuity based on established expectations.

Motivation: The inclination and ability to mobilize the requisite motivation toward the effective discharge of task engagement activities. Balancing the task challenges with task skills throughout work tasks provides individuals with the conditions for and the opportunity to experience behavioral and motivational reinforcement.

Organization: The ability to organize all aspects of a task or activity and to manage task artifacts (e.g., tools, materials, equipment, etc.), on-task behaviors, and associated factors (i.e., self, other, accommodations, and environment) of task planning.

Planning: The ability to plan and schedule tasks from start to finish safely, accurately and to anticipate needs while engaged in the task. Task planning refers to goal-directed behavior including the acquisition, storage, and allocation of information, materials, resources, tools, and equipment. Task planning encompasses all aspects of task engagement.

Production: The ability to optimize on-task behavior (working toward at least a matching or an above average balance between challenges and skills). Task production assumes the intent to do well (i.e., the intention to invest the effort and to persist) leading to behavioral self-reinforcement for the work activity.

Responsibility: The ability to accept responsibility and be accountable for those expectations expressed by an employer. Task responsibility refers to client honesty, trustworthiness, and reliability. This includes the ability to follow laws, work rules, procedures, ethical guidelines, and best practices.

Safety: The ability to determine safe task activities for self and others and to plan and organize around safety guidelines. The well-being of self and others is an overriding focus of task safety.

Time management: Ability to engage in tasks efficiently and in a goal-oriented manner. Activity alone does not necessarily equate with time management if the activity is not effectively achieving task-related goals and achievement expectations. This includes the elimination of task waste (e.g., time, effort, and resources). Focusing on the other common task characteristics has the effect of reducing waste and increasing efficiency.

When we are actively engaged in a work task and demonstrate to the best of our ability (with or without accommodations) the task characteristics listed above, we are demonstrating on-task behavior. Our motivation, when focused on effectively engaging in work activity, reflects having a sense of purpose, building self-reliance and greater independence, feeling connected to others, and working toward achieving the goals we set for ourselves. In short, on-task behavior leads to empowerment. It reinforces motivation and orients us toward successful vocational outcomes. Cobigo, Lachapelle, and Morin define off-task behavior as those activities we engage in which demonstrate other behaviors which are not required to complete a work task (e.g., taking more time than needed to return from break, excessive talking about personal matters while working, waiting to be told what to do when we know what needs to be done) (2010, p. 246). Off-task behavior may be reflective of misaligned motivations for working (e.g., focusing on obtaining a job solely for social engagement or to get out of the house). With off-task behavior many, if not all, of the functional elements suffer—negatively impacting empowerment and self-reliance. Fig. 16.2 illustrates the task activities discussed earlier, each of which may impact on one or all of the accommodation system domains (i.e., a person's situated experience). (Fig. 16.2).

MATCHING PERSON AND TECHNOLOGY

Providers increasingly are better able to respond to consumers' different needs and preferences because the variety of technology and other support options continue to expand. However, the increased availability of AT options has made the process of matching a person with the most appropriate device more complex because people's predisposition to, expectations for, and reactions to technologies and their features are highly individualized and personal. These predispositions, expectations, and reactions emerge from varying needs, abilities, preferences, and past experiences with and exposures to technologies. Predispositions to technology use also depend on adjustment to disability, subjective quality of life or sense of well-being, a person's outlook and goals for future functioning, expectations held by one's self and others, and financial and environmental support for technology use. The Matching Person and Technology (MPT) Model accounts for all of these influences.

In addition to the needs and preferences of the user, a good match of person and technology requires attention to aspects of the environments in which the technology will be used and the various functions and features of the technology. If the match is not a quality one from the standpoint of the consumer, the technology may not be used, or will not be used optimally. Indeed, the overall nonuse or discard rate has been approximately 30% for the past 30 years.

An assessment process exists which has been effective in organizing the influences impacting technology use: the MPT Model and assessment instruments (Scherer, 2005). It consists of instruments which have been validated for use by persons with disabilities ages 15 and up (a measure targeted to technology use by infants and children has also been developed: matching AT and child has an early intervention and a version for K-12 special education students).

The MPT model emerged from grounded theory research. To operationalize the model and theory, an assessment process consisting of the above instruments was developed from the experiences of technology users and nonusers through participatory action research. The development and validation of the MPT assessments followed the recommended steps of professionally approved standards as found in *Standards for Educational and Psychological Testing*: (1) concept definition and clarification, (2) draft of items and response scales, (3) pilot testing, and (4) determination of measure quality and usefulness (Scherer, 1995).

The MPT is a practical and research resource to identify the most appropriate technology for a person in light of the user's needs and goals, barriers that may exist to optimal technology use, areas to target for training for optimal use, and the type of additional support that may enhance use. After the person has received the most appropriate technology for his or her use, the MPT forms are administered at one or more times post-AT acquisition to assess changes in perceived capabilities, quality of life/subjective well-being, and such psychosocial factors as self-esteem, mood, self-determination, and social participation and support.

The *MPT model and accompanying assessment instruments* have been judged by peers in the field as effective in assessing and organizing the many influences which impact on the use of assistive, educational, workplace, healthcare, and general technologies. Ongoing research on the validation of the instruments has shown their utility within and across specific populations of individuals with disabilities. The model has been confirmed by several other AT researchers and authors (e.g., Cook & Hussey, 2008; Lasker & Bedrosian, 2001; Wielandt, Mckenna, Tooth, & Strong, 2006). MPT results have been used clinically and incorporated into AT funding requests and justification reports as well as program evaluations. The MPT Model and assessments have been translated into six languages and have been the focus of eight doctoral dissertations. Both the books *Living in the State of Stuck* and *Matching Person & Technology (MPT) Model Manual and Accompany Assessment Instruments* are used in many academic courses for rehabilitation and allied health professionals and included as a reference text for the Credentialing Examination in Assistive Technology sponsored by the Rehabilitation Engineering and Assistive Technology Society of North America (RESNA). The Assistive Technology in Occupational Therapy self-paced clinical course sponsored by the American Occupational Therapy Association advocates the MPT assessment, "Assistive Technology Device Predisposition Assessment." Of the 59 programs studied in Europe, Canada, and the United States by the European Union-sponsored project, Empowering Users Through Assistive Technology (EUSTAT), the Institute for Matching Person & Technology was selected as one of the top seven "outstanding programs."

Validation studies of the psychometric properties of the MPT assessments have been conducted by independent investigators, a few examples of which are

- The Survey of Technology Use (SOTU) was used by researchers at the University of Rome in Italy while performing, on behalf of the World Health Organization, a standardization of World Health Organization Disability Assessment Schedule (WHODAS). Data from the WHODAS II, the Coping for Stressful Situations, and the SOTU were compared between two matched samples of university students in Italy and in the United States (Federici, Scherer, Micangeli, Lombardo, & Olivetti Belardinelli, 2003).
- To assess the effectiveness of a college course on adapted computer use, 14 college students with disabilities (more than half having complete or partial) eyesight loss identified factors that influenced them to adopt or reject a device for computer access. The results provide evidence of the usefulness of the MPT model and Assistive Technology Device Predisposition Assessment (ATD PA) items as applied to computer access technology for college students (Goodman, Tiene, & Luft, 2002).
- The *initial worksheet for the MPT, SOTU, and ATD PA* were used in research conducted in Italy to validate the utility of BLISS2003, a support for augmentative and alternative communication use (Gatti, Matteucci, & Sbattella, 2004).

- Education researchers used the Educational Technology Predisposition Assessment (ET PA) when preselecting person characteristics relevant to the use of educational technologies (Albaugh & Fayne, 1996; Albaugh, Piazza, & Schlosser, 1997).

Finally, the *Matching Person & Technology (MPT) Model Manual and accompanying assessment instruments* are reviewed in Buros Publications, Tests in Print V and Mental Measurements Yearbook (Scherer, 2001).

SUMMARY

This chapter offered information, tips, tools, and strategies that may be employed as components in the AT service delivery process (for AT and disability-employment professionals alike). There are a variety of case management styles AT and disability-employment professionals will invariable come into contact with so a basic knowledge of these can help them facilitate a continuous flow of services relevant to AT service delivery. Standardizing processes, such as case noting, has a stabilizing effect on AT service delivery and case management which can help reduce programmatic pressures across both formal and informal processes within and across an agency's programs. Reflecting on the nature of employment—the inherent qualities and the expectations they impose on consumers—and effective feature matching as proposed by the Matching People and Technology Model offers disability-employment professionals useful insights into context, activities, and a specific methodology for meeting consumer needs on the job.

REFERENCES

Albaugh, P. R., & Fayne, H. (1996). The ET PA for predicting technology success with learning disabled students: Lessons from a multimedia study. *Technology & Disability*, *5*(4), 313–318.

Albaugh, P. R., Piazza, L., & K., Schlosser (1997). Using a CD-ROM encyclopedia: Interaction of teachers, middle school students, library media specialists, and the technology. *Research in Middle Level Education Quarterly*, *20*(3), 43–55.

Commission on Rehabilitation Counselor Certification. *Code of professional ethics for rehabilitation counselors*. (2017). Retrieved from 〈https://www.crccertification.com/filebin/pdf/ethics/CodeOfEthics_01-01-2017.pdf〉.

Commission on Rehabilitation Counselor Certification. (2018). *CRC/CRCC scope of practice: Selected definitions*. Retrieved from https://www.crccertification.com/crc-crcc-scope-of-practice.

Cook, A. M., & Hussey, S. M. (2008). *Assistive technologies: Principles and practice* (3rd ed.). St. Louis, MO: Mosby.

Federici, S., Scherer, M., Micangeli, A., Lombardo, C., & Olivetti Belardinelli, M. (2003). A cross-cultural analysis of relationships between disability self-evaluation and individual predisposition to use assistive technology. In G. M. Craddock, L. P. McCormack, R. B. Reilly, & H. T. P. Knops (Eds.), *Assistive technology—Shaping the future* (pp. 941–946). Amsterdam: IOS Press.

Gatti, N., Matteucci, M., & Sbattella, L. (2004). An adaptive and predictive environment to support augmentative and alternative communication. In J. Klaus, K. Miesenberger, D. Burger & W. Zagler (Eds.), *Computers helping people with special needs, 9th international conference, ICCHP 2004, Paris, France, July 7–9, 2004, proceedings. Series: Lecture notes in computer science, Vol. 3118* (pp. 983–990). Heidelberg, Germany: Springer-Verlag.

Goodman, G., Tiene, D., & Luft, P. (2002). Adoption of assistive technology for computer access among college students with disabilities. *Disability and Rehabilitation, 24*, 80–92.

Greenwood, R., & Roessler, R. T. (2018). Systematic caseload management. In R. T. Roessler, E. E. Rubin, & P. D. Rumrill (Eds.), *Case management and rehabilitation counseling: Procedures and techniques (5th ed.*, pp. 241–254). Austin, TX: Pro-Ed.

Lasker, J. P., & Bedrosian, J. L. (2001). Promoting acceptance of augmentative and alternative communication by adults with acquired communication disorders. *Augmentative & Alternative Communication, 17*(3), 141–153.

Meeks, J. B. (2001). A social work case management experience in a managed care setting: The need for effective communication. *Home Health Care Management & Practice, 13* (6), 444–451.

Moore, S. (1992). Case management and the integration of services: How service delivery systems shape case management. *Social Work, 37*(5), 418–423.

Mueser, K. T., Bond, G. R., Drake, R. E., & Resnick, S. G. (1998). Models of community care for severe mental illness: A review of research on case management. *Schizophrenia Bulletin, 24*, 37–74.

Rehabilitation Engineering and Assistive Technology Society of North America. *Candidate handbook: Code of ethics.* (2017). Retrieved from ⟨http://www.resna.org/sites/default/files/ATPCIB_2017_02.pdf⟩.

Røhnebæk, M. (2012). Standardized flexibility: The choreography of ICT in standardization of service work. *Culture Unbound: Journal of Current Cultural Research, 4*, 679–698.

Scherer, M. J. (1995). A model of rehabilitation assessment. In L. C. Cushman, & M. J. Scherer (Eds.), *Psychological assessment in medical rehabilitation* (pp. 3–23). Washington, DC: APA Books.

Scherer, M.J. (2001). Test review of the Matching Person & Technology (MPT) Model manual and accompanying assessment instruments. In B. S. Plake & J. C. Impara (Eds.), *The Fourteenth mental measurements yearbook* [electronic version]. Retrieved from the Burros Institute's Mental Measurements Yearbook online database.

Scherer, M. J. (2005). *The Matching Person & Technology (MPT) Model manual and assessments* (5th ed.). Webster, NY: The Institute for Matching Person & Technology, Inc, [CD-ROM].

Shay, A. F., Anderson, C. A., & Matthews, P. (2017). Empowering youth self-definition and identity through assistive technology assessment. *Vocational Evaluation and Work Adjustment Association Journal, 41*(2), 78–88.

Wielandt, T., Mckenna, K., Tooth, L., & Strong, J. (2006). Factors that predict the post-discharge use of recommended assistive technology (AT). *Disability and Rehabilitation: Assistive Technology, 1*(1–2), 29–40.

Woodside., & McClam. (2003). *Generalist case management: A method of human service delivery* (2nd ed.). Pacific Grove, CA: Brookes/Cole.

Wisconsin Division of Vocational Rehabilitation. *Case Noting Style Guidance and Best Practice*. (2015). Retrieved from ⟨https://dwd.wisconsin.gov/dvr/info_ctr/tech/case_noting_style.pdf⟩.

Wong, H. *Improving case management skills for effective vocational rehabilitation services* [PowerPoint Presentation]. (2009, May). Retrieved from ⟨http://slideplayer.com/slide/5988633/⟩.

Wüllenweber, K., Beimborn, D., Weitzel, T., & König, W. (2008). The impact of process standardization on business process outsourcing success. *Information Systems Frontiers, 10*(2), 211−224.

Index

Note: Page numbers followed by "*f*," "*t*," and "*b*" refer to figures, tables, and boxes, respectively.

Printed in the United States
By Bookmasters